Advanced Aerobatics

Other books by Geza Szurovy

Basic Aerobatics by Geza Szurovy and Mike Goulian
Cutting the Cost of Flying
Fly for Less
Learjets by Geza Szurovy (Motorbooks International)
Profitable Photography, Start and Run a Moneymaking Business
Renting and Flying Airplanes Worldwide

Other books in the PRACTICAL FLYING SERIES

Handling In-Flight Emergencies by Jerry A. Eichenberger
Cockpit Resource Management: The Private Pilot's Guide by Thomas P. Turner
The Pilot's Guide to Weather Reports, Forecasts, and Flight Planning 2nd Edition by Terry T. Lankford
Weather Patterns and Phenomena: A Pilot's Guide by Thomas P. Turner
Cross-Country Flying by Jerry A. Eichenberger
Avoiding Mid-Air Collisions by Shari Stamford Krause, Ph.D.
Flying in Adverse Conditions by R. Randall Padfield
Mastering Instrument Flying 2nd Edition by Henry Sollman with Sherwood Harris
Pilot's Avionics Survival Guide by Edward R. Maher
The Pilot's Air Traffic Control Handbook 2nd Edition by Paul E. Illman
Advanced Aircraft Systems by David Lombardo
The Pilot's Radio Communications Handbook 4th Edition by Paul E. Illman
Night Flying by Richard F. Haines and Courtney L. Flatau
Bush Flying by Steven Levi and Jim O'Meara
Understanding Aeronautical Charts 2nd Edition by Terry T. Lankford
Aviator's Guide to Navigation 3rd Edition by Donald J. Clausing
Learning to Fly Helicopters by R. Randall Padfield
ABC's of Safe Flying 3rd Edition by J.R. Williams
Flying VFR in Marginal Weather 3rd Edition by R. Randall Padfield
The Aviator's Guide to Flight Planning by Donald J. Clausing
Better Takeoffs and Landings by Michael C. Love
Aviator's Guide to GPS 2nd Edition by Bill Clarke

Advanced Aerobatics

Geza Szurovy
Michael Goulian

McGraw-Hill
New York   San Francisco   Washington, D.C.   Auckland   Bogotá
Caracas   Lisbon   London   Madrid   Mexico City   Milan
Montreal   New Delhi   San Juan   Singapore
Sydney   Tokyo   Toronto

Library of Congress Cataloging-in-Publication Data
Szurovy, Geza, date.
    Advanced aerobatics / Geza Szurovy, Mike Goulian.
        p.   cm.
    Includes index.
    ISBN 0-07-063302-9 (pbk.).—ISBN 0-07-063303-7 (hard)
    1. Stunt flying. I. Goulian, Mike. II. Title.
TL711.S8S9798   1996
797.5'4—dc20   96-18009
                                                   CIP

McGraw-Hill

A Division of The McGraw·Hill Companies

  7 8 9 10   QSR/QSR   0 9 8 7 6 5

ISBN 0-07-063302-9 (PBK)
ISBN 0-07-063303-7 (HC)

*The sponsoring editor for this book was Shelley Chevalier, and the production supervisor
was Suzanne Rapcavage. It was set in Palatino Roman by Jana Fisher through the services
of Barry E. Brown (Broker—Editing, Design and Production). Unless otherwise noted,
photography by Geza Szurovy.*

Printed and bound by Quebecor Dubuque.

McGraw-Hill books are available at special quantity discounts to use as premiums
and sales promotions, or for use in corporate training programs. For more
information, please write to the Director of Special Sales, Professional Publishing,
McGraw-Hill, Two Penn Plaza, New York, NY 10121-2298. Or contact your local
bookstore.

 This book is printed on recycled, acid-free paper containing a minimum
of 50 percent recycled, de-inked fiber.

For Rick Massegee
in loving memory

DISCLAIMER

All material in this book should be used as a source of general information only. It is the responsibility of every pilot intending to learn aerobatics to receive appropriate comprehensive dual aerobatic instruction from a qualified aerobatic instructor and comply with all regulations and procedures in effect. It is the responsibility of the pilot in command to consult all official sources of information relevant to every aspect of a proposed flight and personally assure compliance with all laws, regulations, and procedures.

Contents

From a garage to the top of the world: Story of a world-class monoplane

10   Advanced spins   **107**

Advanced Aerobatic Maneuvers in History

11 Gyroscopic maneuvers   **123**

World champion Xavier de L' Apparent's four-minute freestyle tips

12   Advanced sequence composition   **131**

Acknowledgments

Advanced aerobatics is intensely competitive, but everyone in the sport to whom we turned for advice responded with an extraordinary level of generosity and encouragement. We would particularly like to thank those colleagues of Mike Goulian who spent a considerable amount of time offering their insightful opinions on the many nuances of advanced aerobatics formed over many years on the competition and airshow circuits. Leo Loudenslager, Patty Wagstaff, Linda Meyers Morrissey, John Morrissey, Phil Knight, Diana Hakala, Randy Gagne, Sergei Boriak, and Xavier de L'Apparent all set aside other commitments without hesitation to help us out.

We would also like to thank Ed Jepsen, who generously lent us his Extra 300 for the in-cockpit photography, Ron Burns for his comments and observations, Tom Hughes for the Sukhoi and Extra graphics, and Shelley Chevalier and Susan Kagey, our editors at McGraw-Hill, for their efforts on our behalf.

Last, a very special thanks to our respective families, Rita and Myron Goulian and Anne Mendenhall, for their remarkable patience and unconditional support for our need to fly.

Geza Szurovy and Mike Goulian
Hanscom Field, Massachusetts

Introduction

The ideal flight crew in the highly automated 21st-century jet's cockpit is said to be a dog and a pilot. The dog is there to bite the pilot if he or she tries to touch anything. The pilot is there to feed the dog. This assertion might be only half tongue-in-cheek at a time when you can fly a Boeing 747 from New York to Tokyo without once touching the controls until you are rolling out on the runway after landing at your destination. It also neatly sums up why so many pilots who want their flying to be truly exciting take up aerobatics.

Pitching your skills against the elements in a DC-3 flying across the Himalayas in instrument conditions using only a compass and a clock to find your way is history. But the adrenaline rush of a snap roll on a vertical down line at full power can be had by all who seek to push their flying skills to the limit instead of passively monitoring computer systems that do the flying for them. If you are the kind of person who takes on the challenge of aerobatics, you are also likely to persevere to the limits of your skills. Our goal is to help you reach those limits.

This volume is a sequel to *Basic Aerobatics* (McGraw-Hill 1994), our first book on the sport. Learning basic aerobatics might initially have been the most demanding task of your flying career. At first the learning curve was steep, but if you were motivated, it went flat relatively quickly. Flying basic maneuvers has become as routine for you as rolling into a steep turn. You hanker for more: snap, vertical, and hesitation rolls; outside maneuvers; rolling circles; tailslides; inverted, accelerated, and flat spins; and gyroscopic maneuvers. In this book you will find them all and more.

You do not have to be a competition pilot to fly advanced aerobatics. Indeed, all pilots with basic aerobatic skills will find it enlightening to explore the realm of advanced maneuvers. As you came to build proficiency in basic aerobatics, it must have been quite a revelation, as it was to us, to realize how little you previously understood about what really happens to an airplane outside straight-and-level flight and moderately banked turns.

Sellers of airplanes and flight training never tire of telling us (and they are right) that nonaerobatic pilots are perfectly safe without ever having to do a spin

Phil Knight in his Extra 300S.

or a full stall during their training, provided they learn to religiously avoid the conditions under which they occur. But to aerobatic pilots that contention is as constraining as saying that children who have not been taught to cross the street are perfectly safe as long as it is effectively drummed into their heads that they are never to leave the block. The whole world waits to be discovered across the street!

Advanced aerobatics can further expand your understanding of an aircraft's behavior throughout its envelope. In time you will realize how much eluded your comprehension when you were flying only the basic maneuvers. But you will also realize something else about aircraft behavior that might surprise you a little. You will find that there isn't always universal agreement among the world's top aerobatic pilots regarding some of the more subtle reasons for an aircraft's behavior under certain extreme flight conditions. The pilots all fly the most complex maneuvers perfectly, but they might argue passionately into the night about some of the more obscure aspects of its causal factors. This is perfectly normal and, once you are able to follow the discussion, quite exciting. At its highest level, advanced aerobatics is a frontier where some subtleties are still not fully understood.

We faced a difficult choice in deciding how technical we should get when explaining the forces acting on an aircraft in complex aerobatic maneuvers. A whole book could be written crammed full of force vectors and lift and drag coefficient formulae. We chose not to do that. In *Basic Aerobatics*, we presented a comprehensive discussion of the fundamental physics at work during aerobatic flight. Those principles apply equally to advanced aerobatics and, in our judgment, are a sufficient foundation for pilots. Any more detailed technical discussion of the physics of aerobatics should be the happy domain of a graduate student of aeronautics who is preferably also an enthusiastic aerobatic pilot.

We do present additional discussion of the forces acting on aerobatic aircraft in those instances in which they are particularly relevant to the advanced aerobatic pilot but of less concern to the basic pilot. We specifically go into greater detail on gyroscopic precession, slipstream effect, torque, sideslip, and the forces at work in autorotation.

While it is not necessary to be a competition pilot to fly advanced aerobatics, you'll find that competition flying is the perfect way to hone your advanced aerobatic skills, and most advanced aerobatic pilots do compete. Competition provides a regular opportunity to have your flying judged. It allows you to observe and benefit from the experiences of your peers. If you are at all competitive, it also lets you experience the thrill of flying to win. All the maneuvers in *Advanced Aerobatics* are taught to be flown to competition standards. Getting started in competition flying, selecting the aerobatic aircraft that best suits your needs, moving up in the ranks, designing advanced sequences, arranging effective coaching and critiquing, and establishing rigorous physical and mental training regimes are also discussed.

Patrick Paris on the low line.

With the exception of advanced spins, learn the maneuvers in the sequence that they are presented in this book. They are in a particular order to teach you what you are ready for at a given skill level. As sequenced, each maneuver plays a part in giving you the foundation for subsequent maneuvers.

When you fly advanced aerobatics, you always run the risk of inadvertently finding yourself in some form of advanced spin. You should determine with your instructor when it is most appropriate for you to get advanced spin training, given the nature of the advanced aerobatic training program designed to meet your personal needs, and you should never fly any of the maneuvers in this book solo without being trained and able to consistently recover from advanced spins.

Speaking of instructors and flying solo, it is also important to point out what this book is not: It is not under any circumstances a self-study manual for learning to fly advanced aerobatics by your lonesome self. Don't ever attempt to fly any of these maneuvers solo without first receiving relevant comprehensive dual instruction from a competent advanced aerobatic instructor in an aircraft with performance and handling characteristics similar to the one in which you fly solo. Contact any of the aerobatic schools in Appendix G, or get a current list from the International Aerobatic Club (the address appears in Appendix A) if you are not already familiar with the aerobatic community.

Advanced aerobatics remains one of the last great opportunities to unleash your physical flying skills to their limits. Unhindered by air traffic control and computers in the cockpit, it is real flying at its primeval best. So strap in, pull vertical, and enjoy it up there!

1

The world of advanced aerobatics

The easy way to define *advanced aerobatics* would be to say that it is everything we didn't cover in *Basic Aerobatics*. But like most sweeping generalizations, that would not be very helpful. The world of advanced aerobatics is vast and complex and does not lend itself to formal definition. To some pilots it might simply mean maneuvers of ever-increasing complexity. To others it might also include advanced-level competition. Even the competition categories can be confusing. Intermediate, Advanced, and Unlimited level competition all count as advanced aerobatics. Only the degree of complexity is different from category to category. In this chapter we outline the main characteristics of the world of advanced aerobatics and help you begin to form an idea about where your goals and aspirations fit in.

WHAT IS ADVANCED AEROBATICS?

There is really no formal dividing line between basic and advanced-level aerobatics. Rather, there is an informal understanding in the aerobatic community that anything beyond simple inside maneuvers and the upright conventional spin is considered advanced aerobatics.

A core collection of individual maneuvers are the building blocks of advanced-level competition aerobatics: the horizontal snap roll, the vertical roll, the hesitation roll, the outside loop, the rolling circle, the tail slide, the vertical snap roll, and the inverted spin. Given the heavy emphasis on autorotational maneuvers in advanced aerobatic competition today, we would have had most of the field covered if we wrote a book called *Snaps and Spins*.

In a category of their own are gyroscopic maneuvers and accelerated and flat spins, flown in the four-minute freestyle event, which is run as an independent encore to a regular competition and is judged separately by its own standards.

Learning to fly the building-block maneuvers well is the key to becoming a good advanced aerobatic pilot, but there is much more to the sport than that. As you become conversant with the maneuvers, you will realize that their variations and combinations are practically infinite along horizontal, vertical, and 45-degree lines of flight (Fig. 1-1). Some even have individual names, such as the humpty bump, but in fact they are variations of a building-block maneuver; a humpty bump is a small half loop on two vertical lines (Fig. 1-2). It may be an outside or inside half loop, entered with a push or a pull.

Fig. 1-1. The inside square loop is really a series of tight quarter loops combined with vertical lines.

Speaking of push and pull, advanced aerobatics has its own jargon. Here are some basic terms we use throughout the book:

inside Any time the aircraft is under positive G, it is said to be inside. In an inside loop, for example, the aircraft is pulling positive G (the pilot's head is pointing toward the inside of the loop).

outside Whenever the aircraft is under negative G, it is said to be outside, as, for example in the outside loop (the pilot's head points away from the inside of the loop).

positive This term is used to describe the aircraft's attitude on the vertical line. On the vertical *up line*, the aircraft is said to be positive if its nose is slightly forward of the vertical line as seen by the judge. On the vertical *down line*, it is said to be positive if the nose is aft of the down line. (See Chapter 4, Figs. 4-1 and 4-2.)

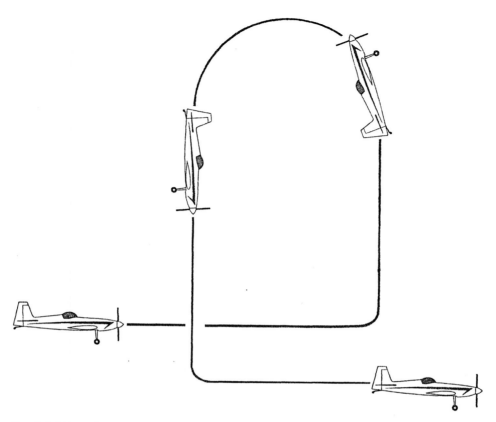

Fig. 1-2. The pull, pull, pull humpty bump consists of two quarter loops and a half loop, combined with vertical lines.

negative This term also describes the aircraft's attitude on the vertical line. On the vertical up line, the aircraft is said to be negative if its nose is slightly aft of the vertical line as seen by the judge. On the vertical down line it is said to be negative if the nose is forward of the down line. (See Chapter 4, Figs. 4-1 and 4-2.)

push To push is to initiate an outside maneuver. The term comes from having to push the stick to apply negative G.

pull To pull is to initiate an inside maneuver. The pilot pulls the stick to apply positive G.

up line The aircraft is said to be on an up line when it is on a straight-line ascending flight path. There are 45° up lines and vertical up lines.

down line The opposite of an up line. The aircraft may be on 45° or vertical down lines.

load To load an aircraft means moving the controls in a manner that applies G to the aircraft. For example, as you pull back the stick to establish the aircraft on the vertical up line, you load the aircraft during the pull.

unload To unload an aircraft means moving the controls in a manner that decreases the G load on an aircraft. For example, as you ease off the stick once the aircraft is established on the vertical, you unload the aircraft.

torque Any movement of the aircraft induced by the engine's torque. This movement might be initiated on purpose, as in a torque roll, or it might be an inadvertent effect, as when the pilot fails to keep torque under control in a hammerhead.

barreled Any type of roll that is not rolling true along the aircraft's longitudinal axis but in a corkscrew fashion around it is said to be barreled.

pinched A loop is pinched when at the end the pilot pulls too hard and the aircraft completes the maneuver on a stunted elliptical path rather than along a perfectly circular one.

flick A flick is another word for *snap roll*, used mostly in British English.

slide An aircraft is said to slide when it descends backward along the vertical up line. This term is also an alternative name for the *tailslide* maneuver.

hunting An aircraft is said to be hunting for the vertical line when it is having difficulty firmly establishing itself on the vertical line.

buried An aircraft is said to be buried when it is too deeply in the stall in a snap roll. The consequence of being buried is to lose too much energy by the time the aircraft completes the maneuver.

You will learn that there are several subtly different ways to accomplish certain maneuvers. Which option you choose is in part a matter of personal preference and in part a function of your aircraft's handling characteristics. We do our best to present all the options and explain our personal preference.

You will also find that different pilots have different preferences for using landmarks and internal timing as references to maintain positional awareness during the maneuvers. These preferences seem to be largely a function of the terrain features in the area where a pilot does most of his or her flying. Pilots used an extensive grid of straight highways and canals, a common terrain feature in Florida, develop a greater instinctive reliance on ground references. Pilots who fly where grid references are largely absent, as is the case in the northeast United States, tend to depend on their internal timing to a greater extent.

THE IMPORTANCE OF BASIC SKILLS

Advanced aerobatics take you closer to the edge than you have ever been before, leaving little margin for error in flying a maneuver well. The slightest deviation from the precise control inputs required can turn an acceptable maneuver into a sloppy mess. Furthermore, to a greater extent than in basic maneuvers, an error at one stage of a maneuver has a compound effect on subsequent stages. At best you'll struggle to make corrections, at worst the whole maneuver might unravel. Precision is paramount in advanced aerobatics.

If one overriding principle can make you the best advanced aerobatic pilot that you can be, it is that you need to have the basic skills down cold. If you don't have this foundation, you'll still be able to slog through the maneuvers and greatly impress the casual observer (and perhaps yourself), but you'll never be a top advanced

Mike Goulian on knife-edge.

aerobatic competition pilot. More importantly, you'll never fly to the limits of your own personal talent. And if something is worth doing, isn't it worth doing well?

The requirement for extreme precision in setting up maneuvers can cause basic aerobatic pilots some difficulty as they make the transition to advanced aerobatics. In basic maneuvers you have more latitude in correcting for such errors as pulling vertical with a wing low. The corrective action might introduce subtle loads that shouldn't be there in the first place, but they don't really matter in a basic maneuver. If, for example, you apply a touch of opposite rudder to correct a wing-low condition on the up line, you can still precisely perform the basic hammerhead in spite of the sideslip you've introduced. However, in an advanced maneuver, such as a vertical roll or a tailslide, any sideslip causes mediocre results at best.

If you've developed any bad habits in basic aerobatics you'll have to unlearn them. Don't let ego get in the way. If a good coach evaluates your performance and tells you that for the next 10 sessions you'll do nothing but learn to fly vertical and 45-degree lines, don't be offended. Be thankful that someone spotted a weakness in your foundation skills and can help make you a much better pilot by working with you to fix it.

ADVANCED COMPETITION FLYING

While you need not be a competition pilot to fly advanced aerobatics, preparing for and flying in competitions is one of the most popular and effective ways of practicing the sport. The competition scene's support structure immeasurably eases your path to becoming a competent advanced aerobatic pilot. The access provided by its network to coaching and peer support is invaluable, as are the opportunities to have your performance regularly evaluated on the competition circuit. We went into some detail in *Basic Aerobatics* on how to get started in competition flying. Here we review the essentials and address issues most relevant to advanced aerobatic pilots.

The International Aerobatic Club and the competition scene

The *International Aerobatic Club* (IAC) is America's national aerobatic organization run by a small professional staff and a large pool of very dedicated volunteers. Its task is to support the sport of aerobatics through a national network of local IAC chapters and the sponsorship and supervision of regional and national aerobatic competitions. The IAC is international in the sense that membership is open to anyone worldwide, and groups of IAC members anywhere are free to form local IAC chapters of their own.

Throughout the year, regional competitions are held in the various parts of the country. Once a year the IAC organizes the U.S. National Championships, traditionally in September. In addition to establishing national champions in the various categories, every other year the nationals are the selection event for the members of the U.S. Aerobatic Team, which competes in the biannual World Aerobatic Championships. The top five men and women finishers in the Unlimited category make the U.S. Aerobatic Team. It is worth noting that in the United States men and women compete and are scored together while in other countries the scores are still segregated into separate men's and women's results.

These aircraft are both CAP 231s. This type swept the 1994 World Championships.

In the United States, there are five competition categories: *Basic, Sportsman, Intermediate, Advanced,* and *Unlimited.* At the national level, events are held in all categories, and at the regional competitions every attempt is made to do so if there are enough entrants.

The Basic and Sportsman categories are the events in which the competitors fly what we consider basic-level aerobatics. The Intermediate category provides the first level of opportunity to fly advanced-level aerobatics in competition. The stakes rise rapidly in the Advanced and Unlimited categories. Take a look at the differences in the level of difficulty in the three categories.

Intermediate

The Intermediate category has very little use of negative-G maneuvers. Negative-G flying is mostly restricted to inverted straight-and-level flight. Snap rolls are featured regularly and partial vertical rolls appear for the first time. Good vertical penetration becomes an important characteristic for aircraft competing in this class.

Advanced

In the Advanced category, negative maneuvers are routinely used for the first time. Greater use is made of more complex variations of maneuvers popular in the Intermediate category. Sophisticated combination maneuvers not seen in Intermediate are also used, and the rolling circle is introduced. Increased use is made of snap rolls.

Unlimited

The Unlimited category places a lot of emphasis on snap rolls and vertical figures. Outside snaps, vertical snaps, and tailslides appear for the first time. Rolling circles are a given. Anything in the FAI (Federation Aeronautique International) Aerobatic Catalog of 15,000-plus figures goes.

In Chapter 11, "Advanced Sequence Composition," we take you through Intermediate, Advanced, and Unlimited sequences to illustrate the differences between them.

In Intermediate, Advanced, and Unlimited competition, pilots compete in the *known, freestyle,* and *unknown* events. The combined scores for all three events determine the final standings. For unlimited pilots, a separately judged and scored four-minute freestyle event takes place at the end of each regular competition.

Known The known event is published for each category for a given year by the relevant supervising body of aerobatic competition (IAC, FAI, and so on). Pilots can train for it in advance as much as they like before they go head to head with their peers at the various contests throughout the year.

Freestyle The freestyle sequence is a sequence composed by each pilot individually for a given competition season. It has to be approved by a judge prior to a competition. Like the known event, it is extensively practiced and rehearsed by the pilot prior to each competition.

A Sukhoi 31 taxiing out to fly its competition sequence.

Unknown   The unknown sequence is composed by the relevant sponsoring entity for each individual contest. Pilots are given the unknown sequence only 18 hours in advance of the event and are not allowed to practice it beforehand. This event tests the pilots' versatility and adaptability. Because of the lack of opportunity to fly the sequence before flying it in competition, it is perhaps the most challenging of the events.

Four-minute freestyle

The four-minute freestyle is a completely separate event from the regular competition, flown in unlimited only, at the end of each regular competition. Participating pilots fly a freestyle sequence of their own composition. Gyroscopic maneuvers are not only allowed, they are expected to be the main elements of the program for pilots hoping to do well. As the name implies, the four-minute freestyle is time-limited to last exactly four minutes. Penalties are imposed for erring on either side of the four-minute limit. It is judged by its own criteria, with heavy emphasis on artistic elements such as originality, versatility, harmony, and rhythm. (See Chapter 10 for more details on the four-minute freestyle.)

At the present time, a competitor in the United States can enter any of the five categories. There are no qualifying requirements. However, if you display such gross ineptitude that you compromise flying safety, the contest officials can bar you from the rest of the competition. This system of relying on the basic human instincts for self-esteem and self-preservation to control who enters what events has worked well. Nobody wants to look like a fool, and nobody wants to get hurt.

You do need an FAI sporting license (Fig. 1-3) to enter IAC-sanctioned Unlimited category events, but this is just a formality. The IAC issues the annually renewable license for a fee.

Fig. 1-3. FAI sporting license.

Given that the choice of participation level is left up to you, it is your responsibility to choose a level that works best for you. The temptation is great to step up to the next level as soon as you can barely struggle through its maneuvers. This strategy is a mistake. You will be so hard-pressed to just get through the sequences in competition that you won't really be able to learn anything from them. You'll be in over your head.

Many pilots are of the opinion that the most efficient way to move up in the ranks is to move up into the next category only when you can consistently finish in the top third of the field in the previous category. We share this opinion and reiterate it throughout this book.

Rules

The rules of competition are set out in the *IAC Official Contest Rules*, an annual publication. It is important to keep current because the rules can and do change from year to year. Aerobatics is constantly evolving, and the rules have to keep pace. The IAC rules are largely based on the guidelines of the Federation Aeronautique International and CIVA, its aerobatic division, which are the ultimate international arbiters of the sport.

The FAI Aerobatic Catalog and the concept of constructing figures

Supplementing the rule books is the *FAI Aerobatic Catalog*. It is the successor document to the *Aresti Catalog*, which was the first comprehensive compilation of aerobatic figures. Rather than go to great length to explain how to read the symbology, we strongly suggest you obtain a copy of the catalog. It contains detailed user's instructions.

Even a cursory look at the *Aerobatic Catalog* reveals the method of constructing aerobatic figures. There are relatively few foundation lines and figures. The rest are variations of the basic figures or composite figures of varying degrees of complexity. Each element has a difficulty coefficient (K). The total of the foundation figure's K and the Ks of the added components defines the total score of the maneuver created by the pilot.

Take, for example, the maneuver in Fig. 1-4. The foundation maneuver is a half loop, worth 6K. It is then loaded with K at both horizontal segments to create a compound maneuver with a high potential score. At the bottom is a four-point roll (11K). At the top is a "3 of 2," which is a 1½ two-point roll (12K), followed by a double inside snap (17K). By the time the maneuver's design is completed, it has gone from 6K to a total maximum score of 46K. It is a single complex maneuver, yet to fly it you need to know how to fly slow rolls, hesitation rolls, half loops, and snap rolls.

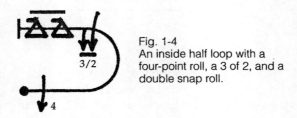

Fig. 1-4
An inside half loop with a four-point roll, a 3 of 2, and a double snap roll.

The Aerobatic Catalog groups the maneuvers into nine families:

1. Lines and angles
2. Turns and rolling turns

3. Combination of lines

4. Spins

5. Hammerheads (stall turns)

6. Tailslides

7. Loops and eights

8. Combination of lines, angles and loops

9. Rolls

Figures from Family 9 must always be combined with a maneuver from another family. Maneuvers in Families 1 through 8 may be flown stand-alone or in combination with elements from Family 9. It is estimated that the total number of figures that can be derived from the catalog exceeds 15,000. If you become proficient in the maneuvers covered in the book, you should be able to handle any combination.

For a flavor of how figures are defined, see Appendix B, which presents selected extracts from the rules. For a comprehensive understanding, obtain the *IAC Official Contest Rules*, as well as the *FAI Aerobatic Catalog*, both available from IAC.

The aerobatic box

The *aerobatic box* is the arena of aerobatic competition. It is an imaginary box of airspace with a base 3300 feet by 3300 feet and a height of 3500 feet agl. The floor of the box varies by competition category. For Basic and Sportsman, it is 1500 feet, for Intermediate it is 1200 feet, for Advanced it is 800 feet, and for Unlimited it is 328 feet, or 100 meters (Fig. 1-5).

The box has two axes, the X and the Y axis. The X axis is the main performance axis in front of the judges, along which most of the maneuvers are flown. The maneuvers flown along the Y axis are known as crossbox or wind-corrector maneuvers used to reposition the aircraft to counter the effects of wind (see Chapters 11 and 13). The box is segmented into three equal zones along the X axis. Each maneuver in a sequence must be flown in the segment of the box in which it is placed on the sequence card.

As you become proficient in flying advanced maneuvers to competition standards, it is important that you learn to fly your sequences in a practice box supervised by a qualified instructor or coach. Many coaches and IAC chapters have a designated FAA-approved practice box. If there are no practice boxes in your area, work with your local chapter and the FAA to establish one.

International competition

Most countries that have an active aerobatic movement have some form of national competition system similar to the U.S. system. There are currently two important international events: the World Aerobatic Championships and the European Championships, held biannually in alternating years. Both the World

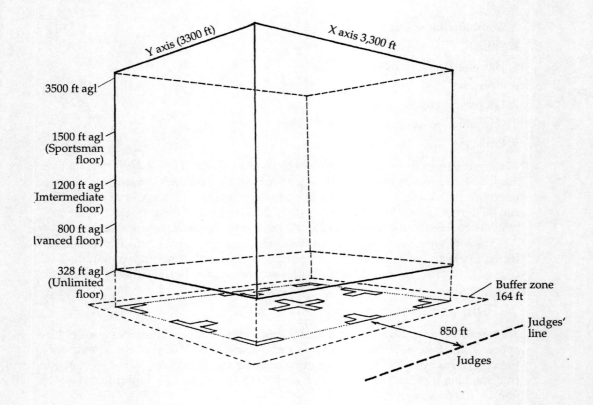

Fig. 1-5 The aerobatic box.

and European Championships are restricted to Unlimited pilots who are members of their national team. Team selection is at the discretion of individual countries. It is possible for foreign nationals to fly in the national championships of countries other than their own on an unofficial basis.

Various efforts are being made to popularize aerobatics and expand opportunities for international competition. There have been experiments with holding world championships for categories other than Unlimited, such as the Advanced World Aerobatic Championships held for the first time in 1994 in South Africa. Attempts are underway to hold an FAI-sanctioned Grand Prix of Aerobatics based on the four-minute freestyle event and consisting of a series of Grand Prix events during the year held at locations throughout the world. The Grand Prix complements the World Aerobatic Championships and has corporate sponsorship. Contestants are selected from the top finishers at the worlds.

The competition scene is never static. In the 1930s it was characterized by freestyle duels at airshows. During the Cold War, East dueled West with massive government support from various countries on both sides. In the entrepreneurial

The Aresti Cup is presented to the world champion. Xavier de L'Apparent (left) was the 1994 recipient. The French coach was Coco Bessier.

1990s, opportunities for corporate sponsorship of events seems to be on the increase. You have to keep your ear to the ground. But you can be sure that whatever the circumstances, the practitioners of the sport will always find a way to compete.

IAC ACHIEVEMENT AWARDS

If you don't want to compete but would like to participate in a systematic program to improve your aerobatic skills, you should consider the IAC's Achievement Awards. It is a program designed along the lines of the various competition categories. Participants fly, at their leisure, a series of maneuvers within a category while observed by an IAC judge. A participant who can successfully complete the required maneuvers within a given category is awarded a certificate of achievement in recognition of the skill level accomplished. Anyone who is an IAC member worldwide can participate. For a full description of this program, see Appendix D.

The kitbuilt Giles 200 (rear) and 202 bring affordable Unlimited aerobatics in a monoplane.

Deciphering Aresti

Today the figures eligible for use in aerobatic competition are found in the *FAI Aerobatic Catalog*. The system of classification and figure construction dates back to the efforts of a colorful Spanish aerobatic pilot, Count Jose Luis Aresti, whose cataloging system became known as the Sistema Aresti and evolved into today's *FAI Aerobatic Catalog*. To this day many pilots use the term *Aresti symbol* for the image of an individual figure.

Each figure within the nine families in the catalog has an assigned numerical value known as K. An inside half loop, for example, is worth 6K. Compound figures are created by combining catalogued figures according to certain simple rules. A compound figure is valued by adding the K of each of its component figures.

The best way to learn aerobatic figure symbology is to buy the FAI catalog and the IAC rule book. To give you a taste of how the system works, we show a few basic concepts and construct and calculate the score of the compound figure used as an example in Chapter 1.

- A solid line represents inside flight, meaning that the aircraft is under positive G.
- A dashed line represents outside flight, meaning that the aircraft is under negative G.
- A small solid circle represents the beginning of a maneuver; a short line perpendicular to the line of flight represents the end of a maneuver.
- Figures are drawn, to the extent possible, to conform to the flight path of the maneuvers they depict.
- Maneuver components difficult to depict by portraying the aircraft's flight path are depicted with a variety of symbols, sometimes aided by numbers, placed on the lines of flight.

Now let's construct a compound figure by combining figures from the various families in the FAI catalog:

1. Let's start with an inside half loop from upright. The catalog number for this figure is 7.2.1, which stands for Family 7, Row 2, Column 1. The catalog tells us it is worth 6K.

Half loop
7.2.1   6k   Constructing a compound aerobatic figure.

2. Now let's add a four-point roll prior to pull up. The number we get from the catalog for this figure is 9.4.3.4: Family 9, 4 point, Row 3, Column 4. The value is 11K.

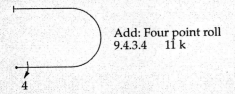

Add: Four point roll
9.4.3.4   11 k

4

3. Now let's add a 1½ two-point roll, called a 3 of 2 in aerobatic terminology, following the looping portion of the figure. Its catalog number is 9.2.3.6: Family 9, 2 point, Row 3, Column 4. It is worth 12K. The small bar below the arrows indicates that the rolls are linked, to be performed without a break.

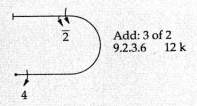

$\overline{2}$

Add: 3 of 2
9.2.3.6   12 k

4

4. Now let's add a double inside snap roll following the 3 of 2. Its number is 9.9.3.8: Family 9, inside snap (9), Row 3, Column 8. The value is 17K.

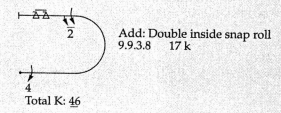

Add: Double inside snap roll
9.9.3.8   17 k

Total K: 46

Our compound figure is complete. All that remains is to add the Ks of its component figures to derive its total value, which is 46K.

2

Getting off the ground

To get the most out of taking the plunge into the world of advanced aerobatics, you need to be well prepared. Although we've said most of what we have to say on the physics of aerobatics in *Basic Aerobatics*, you should review and gain a greater understanding of some theory that is particularly pertinent to advanced maneuvers. You should at least take a first stab at defining your aerobatic goals and objectives and consider them in view of your skills and proficiency. You need to make a decision on the type of aircraft in which you want to train, and you should have a clear understanding of how and where to find good instruction and coaching.

THEORY REVISITED

A handful of phenomena have a greater effect on the more demanding and complex advanced maneuvers than on the basics. Gyroscopic effect, slipstream, and torque effect all manifest themselves regularly. You need to know enough about them to be able to recognize them and deal with them appropriately. To properly set the scene, let's review some very basic points about energy, lift, and angle of attack.

Energy management

The late great Eric Müller, 1974 European aerobatic champion and multiple Swiss national champion, had an elegant way of explaining the role of energy in aerobatics. Throw up a stone vertically, he said, and it starts off with a certain speed and gradually slows until its speed is zero. At this point, a physicist might say that the stone has exchanged its kinetic energy for potential energy. But Eric put it more simply for the benefit of aerobatic pilots: The stone has traded all of its airspeed for altitude.

After a momentary pause the stone starts back down, building speed and arriving back at its starting point at a speed nearly equivalent to its original starting speed (the small difference is due to energy lost to drag). The stone traded its altitude back into airspeed.

A conscious exchange of airspeed for altitude and back again is energy management and is a fundamental concept of performing aerobatic sequences. And remember: As long as you don't run out of airspeed, altitude, and ideas at the same time, you'll be fine.

Lift and angle of attack

In *Basic Aerobatics* we didn't go into what happens to lift once the wing exceeds the critical angle of attack (AOA). The conventional explanation is that lift ceases and the airplane stalls. The airplane does, indeed, enter a mode of flight that we call a stall, but the fact is that some lift continues to be generated even beyond the critical angle of attack. To illustrate the point, in Fig. 2-1 we charted the lift coefficient of a typical symmetrical airfoil for changes in angle of attack. Note that while lift diminishes beyond the critical angle of attack, it continues to be generated to some degree.

It is important to understand this point because it means that even beyond critical angle of attack we can utilize lift to achieve certain objectives by creating lift differentials with the manipulation of the controls. This concept is important in understanding spins and snap rolls and in learning to use this lift to our advantage in advanced spins.

Related to this topic is another relationship that helps refine our understanding of the aircraft's behavior beyond critical angle of attack: the relationship between lift and drag. Figure 2-2 charts a wing's lift and drag coefficients. Look at the values beyond the critical AOA. Lift continues to exist but is diminishing. Drag, however, continues to increase. This, too, is a relationship we put to work in advanced spins.

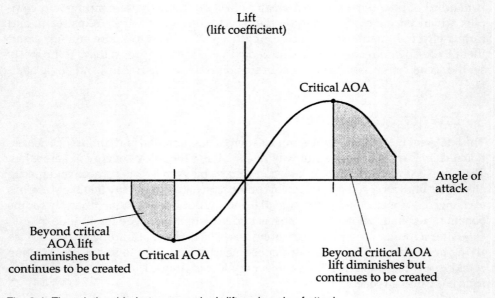

Fig. 2-1. The relationship between a wing's lift and angle of attack.

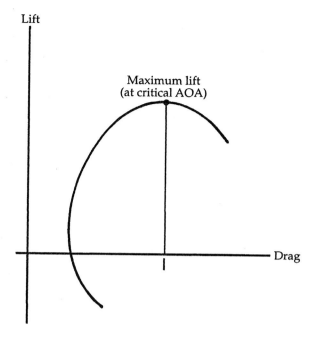

Fig. 2-2. The relationship between a wing's lift and drag. Note that beyond critical AOA, lift diminishes but drag continues to increase.

Gyroscopic precession

The spinning propeller is, in effect, a giant gyroscope. It is a spinning mass. Spinning masses have been observed by physicists to behave in certain predictable ways when subjected to outside forces. The phenomenon is referred to as gyroscopic effect. Aerobatic pilots need to know the consequences of gyroscopic effect to maximize their control over their aircraft.

A spinning mass rotates around a particular axis. (Figure 2-3 shows the three axes of an airplane.) If an outside force is applied to the spinning mass in another axis, the resultant force causes a rotation along an axis 90° ahead ("downstream") of the applied force's axis. To understand how this concept applies to a spinning propeller and affects an aircraft in flight, take a look at Fig. 2-4.

1. The propeller is rotating clockwise (as seen from the pilot's seat) along the A axis, which is the aircraft's longitudinal axis.
2. The pilot applies aft stick and pitches the aircraft up. This, in effect, is an attempt at the rotation of the spinning propeller along the B axis, which is the aircraft's lateral axis.
3. According to the law of gyroscopic effect, the resultant force is on an axis 90° ahead (downstream, in the direction of the propeller's rotation) of the axis on which the outside force (the pitch up) was applied. The C axis is 90° ahead of the B axis. The resultant force, therefore, rotates the spinning propeller around the C axis from left to right. It is also the aircraft's vertical axis. The aircraft's nose moves to the right.

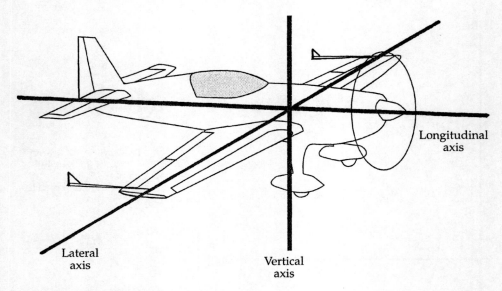

Longitudinal
axis

Lateral
axis

Vertical
axis

Fig. 2-3. The three axes of an aircraft.

If the nose is pitched down the effect is reversed. Gyroscopic precession moves the nose to the left. See if you can reason your way through this effect by using the example in Fig. 2-4.

Gyroscopic effect during pitch up or pitch down is important to counter in abrupt pulls and pushes to or from vertical. We compensate with rudder.

In aerobatic aircraft, gyroscopic effect can also be a factor in flying maneuvers that require an application of yaw. Applying the laws of gyroscopic effect, we can deduct that if the propeller is turning clockwise (to the right) as seen from the pilot's seat, a yaw to the left causes the nose to pitch up. A yaw to the right causes the nose to pitch down.

Gyroscopic effect due to yaw is a factor in flying snap rolls. In an aircraft with a right-turning propeller, a snap roll to the left results in a greater arc inscribed by the nose during the maneuver, because when the aircraft is yawed to the left, gyroscopic effect causes the nose to pitch up. In a snap roll to the right, the nose inscribes a narrower arc around the flight path because gyroscopic effect pushes the nose down as the aircraft is yawed to the right.

The degree of gyroscopic effect varies with the rate at which a given propeller spins. The faster the rpm, the greater the gyroscopic effect. We find this characteristic useful to make spins go flat by bringing the power from idle to full power.

If in flight you are ever in doubt about the effect of gyroscopic precession and want to know what will happen before you pitch up or down or yaw left or right,

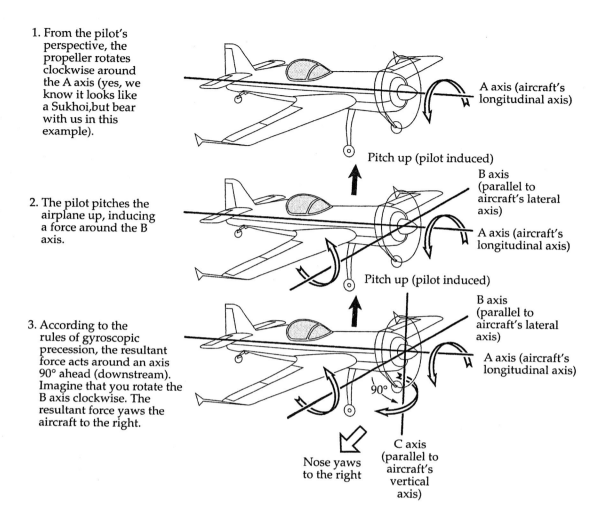

1. From the pilot's perspective, the propeller rotates clockwise around the A axis (yes, we know it looks like a Sukhoi, but bear with us in this example).

A axis (aircraft's longitudinal axis)

Pitch up (pilot induced)

2. The pilot pitches the airplane up, inducing a force around the B axis.

B axis (parallel to aircraft's lateral axis)

A axis (aircraft's longitudinal axis)

Pitch up (pilot induced)

3. According to the rules of gyroscopic precession, the resultant force acts around an axis 90° ahead (downstream). Imagine that you rotate the B axis clockwise. The resultant force yaws the aircraft to the right.

B axis (parallel to aircraft's lateral axis)

A axis (aircraft's longitudinal axis)

90°

Nose yaws to the right

C axis (parallel to aircraft's vertical axis)

Fig. 2-4. The effect of gyroscopic precession.

there is an easier way to remember than trying to mentally visualize the forces (Fig. 2-5). Just remember the word "UPRIGHT" and the logic of action-reaction. The action is UP, as in "pitch up," and the reaction is RIGHT, as in "nose right." You can then easily remember that the opposite happens if you pitch down.

Remember "RIGHTDOWN" for recalling what happens in yaw. The action is RIGHT, as in "yaw right," and the reaction is DOWN, as in "nose down."

Another reminder is a hand clue. Make a fist with your right hand with your forefinger and thumb extended. Lift your hand up to your face so that the nails on your clenched fingers face you. When your forefinger is pointing UP, in the direc-

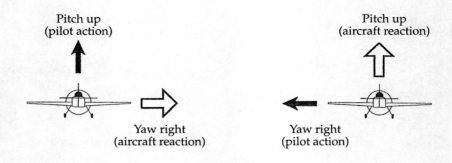

Fig. 2-5. Gyroscopic precession—what the pilot has to remember.

tion of pitch, your thumb will be pointing RIGHT, the direction the nose will go due to gyroscopic precession.

Of course, if you forget all this you can just build up a big head of steam, do a high G pull, and see which way the nose goes. And if you still have trouble remembering when you are roaring along the X axis in a competition, you either haven't trained enough or maybe you should consider taking up fly fishing.

Slipstream

The turning propeller sets off a spiraling column of air that trails behind it in a corkscrew fashion (Fig. 2-6). The vertical stabilizer in the path of the corkscrewing air particles (and to some extent also the horizontal stabilizer) is being pushed in a turning, twisting motion as it is struck by the particles. The net effect of slipstream is that it wants to push the tail of the aircraft to the right (on an airplane with a right-turning propeller), yawing the airplane to the left.

At high speeds, slipstream is not a problem because the air particles are spread out, trailing behind the aircraft on such a long trajectory that their effect is mini-

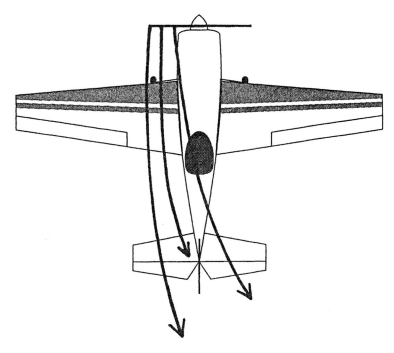

Fig. 2-6. Slipstream effect.

mal. However, at high power and near-zero airspeed, a condition generally encountered only by aerobatic aircraft at the top of a vertical line, primarily in the hammerhead, the particles are packed densely together, making slipstream effect much more pronounced and requiring corrective action with opposite rudder.

In classic aerobatic maneuvers, such as the hammerhead, slipstream effect is undesirable. In gyroscopic tumbling maneuvers, however, it is an ally. Setting up a tumble in which gyroscopic precession and slipstream effect act in the same direction yields the most spectacular results.

Torque

For every action there is an equal and opposite reaction. So says Newton's Third Law of Motion, one of the most elementary rules of physics. When we talk about torque effect in an airplane, we are referring to this law at work as the propeller turns. Torque effect is the tendency of the airframe to rotate in the direction opposite to the propeller's rotation in reaction to the force of the propeller's rotation. Generally the aircraft's mass, aerodynamic forces, and trim are sufficient to keep torque effect in check. In all propeller-driven aircraft, the engine is mounted at an offset to the right and down, optimized to cancel out torque effect at cruise speed. But at high power and low or near-zero airspeeds, torque overcomes its opposing forces and the fuselage starts rotating in reaction. It is countered with opposite aileron.

Torque becomes a factor as the aircraft slows or comes to a stop on the vertical line. Torque effect is also noticeable in snap rolls. Most aircraft with a right-turning propeller snap faster to the left than to the right because of torque effect.

Torque effect is greater the more powerful the engine and the heavier the propeller. Torque was a particularly big issue for the powerful World War II fighters, which were equipped with engines in excess of 1800 horsepower (hp). If the pilot of a P-51 Mustang was foolish enough to take off at full power, the airplane would roll over on its back the minute the wheels left the ground, and the pilot could do nothing about it.

PILOT PROFICIENCY AND CONDITION

You should be a competent basic aerobatic pilot before you move on to advanced maneuvers. You need not be extraordinarily talented to do well at advanced aerobatics. What you need most is a strong drive to fly advanced aerobatics, which usually comes from a capacity to genuinely enjoy it. If you can't get enough of basic aerobatics, you are a prime candidate for advanced aerobatics. But if you learned basic aerobatics in an effort to become a better pilot rather than with the intention of flying aerobatics regularly, you should also consider a similar experience with advanced aerobatics. You should be able to handle it, and it will teach you that much more about flying and yourself.

As you up the ante in advanced aerobatics, you will be subjecting yourself to some pretty hefty G loads. High negative Gs are a whole new (and, for some, not a particularly pleasant) experience. It is therefore useful to be in good physical condition. If you are generally unaffected by the rigors of basic aerobatics, you should do fine in the elementary phase of advanced aerobatics.

Besides being generally fit, you should always build up gradually to the more punishing loads of advanced maneuvers. Many advanced maneuvers can be performed with no more strain than basic maneuvers. Start exceeding your basic limits in small increments, and see how it goes. Bear in mind that even the top-level competition pilots build up G tolerance gradually after a break of any length from advanced-level aerobatics.

As you build experience and want to do well at the uppermost levels of unlimited competition, you should be supremely fit, no doubt about it. Later in this book we address the issue of physical training for competition flying. But at the outset you are a long way from the nationals and the worlds and can work on improving your physical condition to parallel the progress of your advanced aerobatic skills.

For more information on the effect of aerobatics on the human body, including discussions of gray-out, black-out, red-out, and G-LOC (G induced loss of consciousness), refer to *Basic Aerobatics*.

AIRCRAFT CONSIDERATIONS AND CHOICE

Many factors go into selecting an aircraft for advanced aerobatic flying. Equally important considerations are your aerobatic objectives, pilot proficiency, aircraft performance, aircraft price, and operating costs. But whatever your choice, always follow the flight manual's directives for flying maneuvers, and never do a maneuver for which your aircraft is not approved by its manufacturer.

The novice is always concerned with the transition from lower performance aircraft that are excellent basic aerobatic trainers, such as the Bellanca Decathlon, into aircraft capable of advanced aerobatics. There is a big jump in performance as you step up, and handling characteristics also become more demanding. The Pitts Specials, the Laser derivatives, and the modern 300-hp monoplanes are all a handful if all your experience is in a Decathlon.

The key to transitioning into high-performance advanced aerobatics is first obtaining appropriate dual instruction in an advanced two-seater. Ever since 1971 the Pitts S-2 series aircraft has been the workhorse of advanced dual instruction. If you can learn to confidently handle the 200-hp Pitts S-2A or its more powerful sibling, the 260-hp S-2B, you can fly any advanced aerobatic aircraft. Many pilots elect to start their aerobatic training outright in the two-seat Pitts. Another good two-seat choice that has been around for years is the French CAP 10, but it never became very popular in the United States, perhaps because of its high price.

If your goal is to fly a 300-plus-hp monoplane, there are opportunities to receive dual instruction in 300-hp two-seaters of equivalent performance to prepare you for the challenge. The best options are the Extra 300 and the Extra 300L, because they are both certified and can therefore be operated by aerobatic flying schools to provide dual instruction. Another excellent two-seater in the 300-hp category is the Sukhoi 29, but in the United States you have to buy one to receive dual

Greg Ryan

The certified Extra 300 two-seat unlimited monoplane, predecessor to the Extra 300L.

Patty Wagstaff waiting for her turn at the world championships in her Extra 300S.

in it because at the moment it does not have its American certification. A limited number of outstanding experimental Staudacher S600s are also around.

An especially encouraging development is the recent appearance of several single-seat and two-seat advanced aerobatic aircraft with engines in the 200-hp-plus range that fit neatly in the niche between the Decathlon and the top unlimited monoplanes.

Recognizing the need for a more affordable option to the 300-hp monoplanes, Walter Extra has introduced the 200-hp two-seat Extra 200. It looks like a slightly scaled-down Extra 300L and provides traditional Extra performance and quality on the economy of the four-cylinder 200-hp engine. Being a certified production aircraft, it should be a popular choice for advanced aerobatic schools. It also comes with an in-

The Sukhoi 29 is also a formidable unlimited two-seat training and competition aircraft.

terchangeable optional canopy that instantly converts it into a "single-seat" airplane.

A number of recently developed homebuilt advanced-aerobatic monoplanes in the 200-hp-plus range are available in kit form, which saves a considerable expense in labor costs for pilots with homebuilding skills. Dan Rhin's single-seat One Design, its two-seat version, the Giles G-200 single seat, and the G-202 two-seat aircraft all exhibit excellent handling characteristics. These aircraft are quite easy to build, and, as more are constructed over the years, they should become more widely available to nonbuilders on the secondhand market.

We should make something perfectly clear about the biplane and the monoplane. The choice between the two is entirely a matter of personal preference. Biplanes and monoplanes of equal performance are capable of doing equally well against one another. The fact that the monoplanes have been edging out the Pitts Specials at the highest level of unlimited competition is because the newer monoplanes are more advanced, higher-performance aircraft. Regarding judging, the current wisdom is that the monoplanes, with their cleaner and more angular lines, display better and are therefore easier to judge when they fly well. However, mistakes are also more visible.

As we go to press, the biplane is about to make an attempt to reclaim its preeminent position in the lineup of top aerobatic aircraft. AVIAT is introducing the Pitts S1-11B, a redesigned Pitts airframe mated to a 300-hp Lycoming IO-540 engine optimized by Monty Barrett.

Fig. 2-10. Biplane and monoplane: The Pitts S-1T and the Zivko Edge 540.

Let's take a closer look at some of the considerations that should help you in choosing the advanced aerobatic aircraft that best meets your needs and objectives.

Aerobatic objectives

You need to have a clear idea of where you see your aerobatic flying taking you. Do you primarily want to learn advanced aerobatics for the experience, to make you a better pilot? Do you want to compete? If you want to compete, do you want to do it at the regional level, or do you have your sights set on the nationals? Is it important to you where you finish in competition, or do you just want to participate and find acceptable any performance that doesn't make a fool out of you?

Perhaps the best way to approach these issues is to ask yourself how intensely you want to get into advanced aerobatics. Your answer will have a lot to do with your choice of aircraft. If you know up front that you are aggressively aiming to do well in unlimited national competition, consider initial advanced aerobatic training in a 200-plus-hp two-seater such as the Pitts S-2B or the Extra 200 as an interim step into a 300-hp single seater. Do learn to perform competently in the lower performance aircraft; otherwise the bigger machine might prove to be too much of a handful.

If you are committed at a lesser level, an aircraft in the 200-plus-hp range is the perfect choice for dual training as well as a single seater.

Pilot proficiency

Don't make the mistake of thinking that because you have been flying high-performance conventional aircraft, such as a Bonanza, Malibu, or some truly hot rods courtesy of Uncle Sam, you are qualified to casually hop into a top 300-hp aerobatic monoplane. Advanced aerobatic airplanes are different animals deserving of a lot of respect and calling for a gradual approach, regardless of your previous experience.

Different pilots make progress in new airplanes at different rates. The only way to find out how rapidly you will take to a new mount is through rigorous dual evaluation and training. And whatever you do, don't rush it.

Aircraft performance

If funds are no object, how much performance you opt for is a personal choice. In broad terms, performance is a function of power and aerobatic handling characteristics. Here are some important performance elements you should consider:

- *Vertical penetration.* Given the emphasis on vertical maneuvers in advanced aerobatic competition, this characteristic is especially important if you want to become a serious competitor. Evaluate climb rate and the maximum distance covered on the vertical line from a pull to vertical at full power at 1000 feet.

- *Roll rate.* This characteristic is another important consideration for competition flying. A Pitts S-2B has a roll rate of 180° per second. A top 300-hp monoplane rolls in excess of 360° per second.
- *Handling characteristics.* Handling characteristics have to do with how light and harmonious the controls are throughout the performance envelope. Some high-performance aerobatic aircraft, such as the Zlin 50, have peculiarly quirky control imbalances that require getting used to. Others can make you think after a few flights that the wings are an extension of your thoughts.

Interestingly enough, among the top unlimited aerobatic aircraft, you'll find little difference in performance. Reputations for projecting brute power, being dainty, or having the best all-around characteristics are perceptions largely in the eye of the beholder. The fact is that all of these aircraft are within a hair of each other in performance. The difference is in the pilot. Ultimately your choice should depend on which aircraft clicks with you. Fly them all, compare them carefully, and see for yourself.

Aircraft price and operating cost

Aircraft price and operating costs can be a limiting factor for many of us in choosing an advanced aerobatic aircraft. The highest performance aircraft are very expensive, as much as 30 to 40 percent above the price of less-competitive, lesser-performing aircraft. The demand for these specialized aircraft is low, so unit costs are high. Their operating costs are also higher, primarily because of their big, thirsty, expensive engines. Also be aware that most propellers used on advanced aerobatic aircraft are distressingly expensive and need costly overhauls at more frequent intervals than the ones bolted to the front end of nonaerobatic aircraft.

You can become just as competent an advanced aerobatic pilot in the lower-performance advanced competition aircraft as in the highest performance ones. The only disadvantage you might encounter because of a performance difference will be at the upper echelons of unlimited competition. Don't worry if you can't afford top of the line. Become the best pilot you can be in a more modest aircraft. By the time you work your way into unlimited competition and become good enough for the differences in aircraft performance to really matter, you'll find a way to afford the best.

In the right hands, Pitts Specials continue to perform strongly in all but the top of unlimited competition, and they are much more affordable than the top monoplanes. The new entries into the 200-plus-hp field also cost considerably less and are almost on par in performance. Other options on the used market are the Extra 230, the Lasers, the Zlin 50, and the Christen Eagle.

If you don't want to be a serious competitor, the most economical option for acquiring an aircraft that is perfectly capable of advanced aerobatics is to buy one that is no longer considered competitive beyond intermediate. A good example is the Zlin 526, the last in the line that pioneered many of the maneuvers flown today.

THE IMPORTANCE OF ADVANCED
INSTRUCTION AND COACHING

Once you become skilled at flying basic aerobatics, it is tempting to think that you can begin to work on flying increasingly advanced maneuvers on your own. To attempt to do so would be a dangerous mistake. Advanced aerobatic maneuvers are highly specialized, leaving little margin for error. Attempting to teach yourself advanced aerobatics would be as foolish as attempting to teach yourself instrument flying without an instructor. Please take to heart our insistence that you learn advanced aerobatics by receiving thorough dual instruction from a qualified advanced-aerobatic instructor. It is the only way to keep your flying safe.

The student of advanced aerobatics needs to accomplish two objectives:

- Learn to fly the maneuvers safely and with confidence. To do so, you must receive dual instruction.

- Learn to fly the maneuvers to competition standards. This second phase of learning is best accomplished with the help of a coach who has experience preparing pilots for competition flying. Coaching is far more important in making you a better advanced aerobatic pilot than you might think. If you don't know what you don't know, how can you move forward? If your objective is to fly to competition standards, reliance on the assistance of a good coach is the only way to reach the next level (we say more about this later). One hour of good coaching far outweighs the value of countless hours of trying to improve your performance by yourself.

Where can you find competent instructors and coaches? The IAC maintains a list of aerobatic schools, listing the aircraft and levels of training available. This list is a good start. Aerobatic schools and coaches also advertise in the IAC's monthly magazine, *Sport Aerobatics*, indicating the level of training offered. However, since the IAC is not in a position to express an opinion on the schools on its list and in the magazine advertisements, it is up to you to do your homework in evaluating the options.

Ideally, your instructor should have a lot of experience teaching advanced aerobatics, but equally important is his or her advanced competition experience. Do not hesitate to ask detailed questions about the instructor's experience. Most instructors are as proud of their achievements in competition as they are of their ability to instruct and will be forthcoming with background information. Also ask around for advice on instruction at the local IAC chapter and from experienced aerobatic pilots.

It is a good idea to take concentrated doses of advanced aerobatic instruction. Setting aside a week in which you can fly twice a day every day is an excellent way to learn advanced aerobatics. Too long an interval between lessons prevents you from building the cumulative experience required to make rapid progress. If you choose to take concentrated instruction, you can also more easily plan to attend an aerobatic school not in your immediate area.

Relatively few people specialize in coaching, and by the time you have a need for them, you'll be quite aware of your options through networking within the aerobatic community. We go into some detail in Chapter 14 on the role of a good coach. A coach will rarely, if ever, fly with you. His or her job is to watch your flying from the judge's line and tell you how to improve it. Ideally, your coach should have considerable competition experience at the level to which you are being coached. He or she should not only be able to evaluate your flying, but thoroughly understand and convey to you what needs to be done in the cockpit to improve your performance.

Receiving coaching in a concentrated dose is even more important than receiving concentrated dual instruction. Given the many subtle nuances and refinements you have to learn and practice to improve your competition performance, anything less than three straight days of coaching is unlikely to be productive. For best results, hire a coach for a week if you can. Coaches usually charge for their services on a per-day basis. Small groups of pilots can hire a coach jointly and get a discount on the daily rate without jeopardizing the individualized attention that is one of the most important aspects of effective coaching.

AND BEAR IN MIND

As we build experience with any particular form of flying and it begins to become routine, some of us are tempted to take for granted certain very important aspects that don't directly involve manipulating the controls. When, for example, have you last boned up on changes to FAR Part 91? Here are a few points to bear in mind now that you are a competent basic aerobatic pilot about to move on to advanced aerobatics.

Federal Aviation Regulations

Keep current on the FARs in general and the clauses that pertain to aerobatics in particular. Part 91 of the FARs spells out limitations on aerobatic flight and the requirement for using parachutes. Rather than quoting them here, we leave it you to check the current regs, since they might change from time to time. Bear in mind and observe the FARs' altitude restrictions in effect for aerobatics. Without a special waiver for low-level airshow flying, these restrictions may be disregarded only in an FAA-approved active aerobatic box.

Your parachute

Never fly aerobatics without a parachute, even though in the United States it is legal to do so if you are flying solo. And give your parachute the respect a piece of equipment that might someday save your life deserves. Be sure to have it repacked at the required intervals. Keep it in its carrying case in a dry, cool place when not in use, and minimize its exposure to the harmful UV rays of direct sunlight. This might sound obvious, but an amazing number of people use their parachutes for everything from pillows when they sunbathe to doorstops in the hangar.

Preflight inspections

Sometimes it can be tempting to do a cursory preflight inspection. After all, haven't you been flying the airplane every other day for most of the summer? But cutting corners can ruin your day. What if, while you were away for a moment, someone surreptitiously lifted the canopy for a peek inside and a pen inadvertently slipped out of his pocket onto the cockpit floor as he leaned over for a better look at the radio? What if a hairline crack appeared on a critical surface since the last flight? What if the slightest oil leak is trying to tell you that the prop seal is about to give up the ghost? What if, what if? All these important items can be easily missed if you become lax about preflights.

Technical inspections at competitions

Don't forget that your aircraft has to pass a technical inspection at every competition before it is allowed to participate. Make sure that it is ready. Practice-day morning is no time to start chasing spare tires because the inspectors won't let you fly with the ones you've been meaning to change for weeks. You should obtain the technical inspection guidelines from the IAC for guidance.

Don't forget to smell the roses

Above all, aerobatics is supposed to be fun! Sometimes we might have to remind ourselves of that important fact. It is especially useful to do so when we become frustrated by feeling that we are not doing as well as we think we should or when learning some ornery maneuver starts feeling like a job rather than a pastime. By all means, give vent to your competitive urges, but go at your own pace and never lose sight of enjoying yourself.

So much for the preliminaries. If you want more detailed information on the physics of aerobatics, the effects of aerobatics on the human body, buying aerobatic aircraft, the IAC, and getting started in competition, review *Basic Aerobatics*. Otherwise, it's time to go flying!

This 300-hp Pitts S-1-11B, flown by Bob Armstrong of the U.S. Aerobatic Team, challenges the top monoplanes.

3

Horizontal inside snap rolls

The horizontal inside (upright) snap roll is essentially a spin on a horizontal flight path. It is an autorotational maneuver. As in a spin, during the snap roll autorotation is induced by differences in the drag and lift coefficients of the wing inside and outside the direction of rotation.

The horizontal snap roll is the ideal initial step beyond basic aerobatics because it is the best way to introduce students to the high roll rates inherent in advanced aerobatics. The aircraft's roll rate in the snap roll is considerably higher than the roll rates in such basic maneuvers as the aileron roll and slow roll. However, the aircraft's flight path in the horizontal plane relative to the ground makes it fairly easy for aspiring advanced aerobatic pilots to learn to maintain spatial orientation. It is also a relatively low-G maneuver (once the aircraft becomes unloaded following an instant of high G as the accelerated stall is induced), placing little strain on the human body.

To initiate a snap roll, the pilot first pitches up the aircraft to exceed critical AOA, as in a spin. There is, however, one important difference. In the snap roll, the pilot stalls the aircraft (exceeds critical AOA) with an abrupt aft stick movement at approximately 1.7 Vso. This, as we discussed in detail in *Basic Aerobatics*, is an accelerated stall. A fraction of a second after critical AOA is exceeded, the pilot applies full rudder in the desired direction of rotation, which causes the aircraft to yaw in the direction of rotation along its vertical axis. As a result of the yaw, the inside wing decelerates and the outside wing accelerates. This causes a decrease in lift generated by the inside (decelerating) wing and a corresponding increase in the lift being generated by the outside (accelerating) wing, inducing the roll component of autorotation (along the longitudinal axis). Remember, some lift continues to exist even beyond critical AOA, and differences in the lift coefficient between the left and right wings result in a roll even if both wings are beyond critical AOA.

The dropping wing's angle of attack (AOA) increases, and the rising outside wing's AOA correspondingly decreases due to the shift in relative wind striking the two wings (although both wings remain above critical AOA, i.e. "stalled"). This further increases the difference in lift coefficient between the two wings. Simultaneously, due to the higher AOA, the inside wing's drag coefficient increases compared to the outside wing, further increasing yaw (rotation) along the vertical axis. The differences in lift coefficients and drag coefficients cause autorotation as the aircraft's flight path continues to proceed along the horizontal plane.

Once autorotation is induced in the snap roll, the pilot takes two other steps that differ from the conventional spin, with the objective of getting through the snap roll as quickly as possible with a minimum loss of airspeed (energy). The pilot accelerates the rotation by slight aileron input in the direction of rotation and follows with the application of slight forward stick. The aileron input accelerates the rotation by increasing the roll rate, and forward stick reduces the aircraft's frontal area exposed to the flight path, resulting in less planform drag, as well as increased rotation (Fig. 3-1). This technique is known in aerobatic terms as acceler-

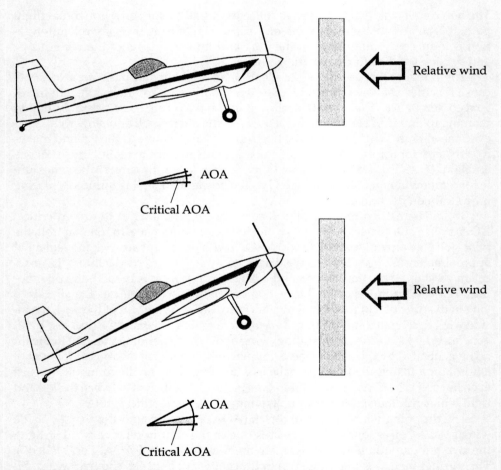

Fig. 3-1. Planform drag—burying the aircraft in the snap roll.

ating the snap roll. A good analogy for this latter point is the effect figure skaters experience in a pirouette when they draw in their arms from overhead to their chest. The repositioning of their arms reduces drag and concentrates the center of mass, significantly increasing the pirouette's rate of rotation.

FLYING IT

Flying the snap roll is fairly easy, but flying it well to competition standards takes a great deal of practice. Let's see how it is done (Fig. 3-2).

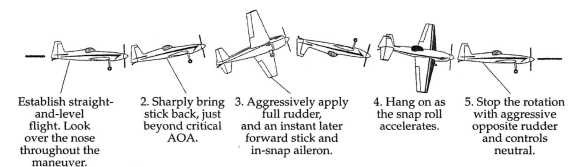

Establish straight-and-level flight. Look over the nose throughout the maneuver.

2. Sharply bring stick back, just beyond critical AOA.

3. Aggressively apply full rudder, and an instant later forward stick and in-snap aileron.

4. Hang on as the snap roll accelerates.

5. Stop the rotation with aggressive opposite rudder and controls neutral.

Fig. 3-2. The horizontal inside snap roll.

1. Establish straight-and-level flight at around 120 to 130 mph in a high-performance monoplane or about 110 to 120 mph in a Pitts or equivalent aircraft. Check your aircraft manual for the correct entry speed. Too high a speed will damage the aircraft. Look over the nose at an easily identifiable reference point straight ahead on the horizon. Since the power setting remains constant during the maneuver, both of your hands will be free to hold the stick. Depending on the aircraft you fly, you might find it more comfortable to perform the snap roll with both hands on the stick because of the stick forces involved. Experiment to see what works best for you. Most Russian competition pilots grab the stick with both hands; most Americans and French prefer the single-handed approach.

2. Once established in straight-and-level flight at the correct entry speed, sharply bring the stick back to raise the nose just beyond critical angle of attack (AOA). The angle above the horizon at which this is accomplished in your particular aircraft should be established by repetitive experimentation. Typically, critical AOA is reached when the nose reaches an attitude of about 18° AOA. It is not necessary to bring the stick full aft to accomplish this condition. How far back the stick comes is determined by your airspeed and how aggressive you are on the elevator. The faster you are, the less aft stick you need (to counter the wing's lift vector), but the faster the movement of the stick needs to be to break the relative wind.

You have to be careful not to go too far beyond critical AOA. If you do exceed critical AOA by a big margin, the aircraft's planform drag will be too high. As a result, the rotation in the snap will be slow, and you'll lose too much airspeed by the time you exit the maneuver. In aerobatic lingo, the aircraft will be "buried" (see Fig. 3-1).

3. As you reach about 50 percent of the pitch up above the horizon toward critical AOA, aggressively apply full rudder in the desired direction of rotation. The idea is to have full rudder deflection as critical AOA is achieved. At this instant, autorotation commences.

4. Once the aircraft starts to autorotate (a split second after you apply full rudder) perform the following control inputs simultaneously:
   - Add forward stick to the point where the angle of attack is just on the edge of the critical angle of attack (the aircraft is just on the edge of flying).
   - Add a very slight touch of aileron in the direction of rotation. The amount of aileron required varies from aircraft to aircraft.

   These control inputs accelerate the snap roll. Forward stick reduces planform drag by reducing the radius of the nose movement around the horizon. Aileron in the direction of rotation increases the lift coefficient on the outer wing, thus accelerating the rate of rotation. You must be careful not to apply too much aileron, because you can reach a point where the increasing drag caused by the down aileron of the outer wing outweighs the benefit of the increased lift coefficient. The outboard aileron will begin to act as a spoiler.

5. Hold full rudder until the instant recovery is desired, which is when the aircraft reaches the attitude it had upon commencing the snap roll. At this point, forcefully apply opposite rudder until the rotation stops (you initiated autorotation with rudder, therefore you have to stop it with rudder). Neutralize rudder and stick, and you'll be flying straight and level with the nose nailed to the reference point on the horizon that you chose to fly the maneuver.

Remember that you put the aircraft into an accelerated stall to commence the maneuver. Being in an accelerated stall, it continued to move along a horizontal flight path while it was autorotating. Therefore, when you stop the autorotation and reduce AOA below critical AOA, you will find yourself in stabilized straight-and-level flight at an airspeed only slightly lower than the entry speed.

THE FINER POINTS

All aerobatic aircraft have slightly different snap-roll characteristics due to differences in aerodynamic traits and weight distribution. To fly snap rolls consistently to competition standards, you have to become intimately familiar with the particular traits of your own aircraft by trial and error. The snap roll is a maneuver in which a little good coaching can be particularly effective, saving you countless hours of flight time that you would spend experimenting on your own trying to refine the basics you learned during dual instruction.

The most significant divergence in the snap roll characteristics of different aircraft stems from the effect of gyroscopic precession. Aircraft that are strongly affected by gyroscopic precession should be snapped only in the direction of the propeller's rotation. If the snap roll is done in the direction of the propeller's rotation, gyroscopic precession tends to push the nose down and toward the center of rotation, reducing the radius of rotation. Remember, the nose is already pitched up past critical AOA, presenting an increased frontal area to the relative wind, resulting in increased drag. Pushing the nose back down reduces the exposed frontal area, reducing drag and resulting in minimal energy loss and a cleaner snap, evidenced by a higher exit speed at the end of the maneuver.

If the snap roll is done against the direction of the propeller's rotation, gyroscopic precession tends to pitch the nose up. This maximizes the aircraft's frontal exposure to the relative wind during rotation (remember, the nose is already just above critical AOA at the commencement of autorotation), resulting in increased drag and a greater loss of energy, evidenced by a greater loss of airspeed upon completion of the maneuver. The slower exit speed makes it more difficult to recover precisely, due to less airflow over the rudder, and leaves less energy for the upcoming maneuvers. However, the rate of rotation is faster in the snap roll performed opposite to the direction of the propeller's rotation, basically due to slipstream and torque (see Chapter 2).

The larger monoplanes with composite propellers and six-cylinder engines tend to be affected less by gyroscopic precession and snap with approximately equal energy loss in both directions. However, because of torque and slipstream effect, both of which induce a general tendency to rotate against the propeller's direction of rotation (see Chapter 1), these aircraft snap roll at a faster rate opposite the propeller's rotation.

COMMON ERRORS

For the aspiring advanced aerobatic pilot, the big challenge is getting accustomed to the rapid roll rate. The high speed at which events unfold initially tends to overload the average student's mind. Additional sources of difficulty are the high control forces and the requirement to induce a rolling maneuver with rudder instead of aileron as the student has been accustomed to doing until encountering the snap roll.

Several techniques are available to make learning the snap roll a smooth, constructive experience. The best way to start doing snap rolls is to do them from lower entry speeds. If you start out with 90 to 100 mph entry speeds, the stick forces are much lower, giving you a feel for the required control inputs without having to fight the airflow pressure on the control surfaces. At these entry speeds, the rotation rate is also slower, allowing you to visually absorb the maneuver without getting disoriented.

Another useful learning technique is to leave out the acceleration of the snap roll until you are comfortable with the basic autorotation. Just apply aft stick briskly, apply the rudder at critical AOA, and watch the rotation rate until the horizon is level again with the sky above.

Most students have to concentrate hard on learning to recover with rudder instead of attempting to do so with aileron. Just think "spin" before each snap roll until recovery with rudder becomes instinctive.

Jim Holland, a well-known aerobatic instructor, has an interesting technique for teaching his students that snap rolls are all about rudder:

1. Reduce the power setting to 50 percent.
2. Raise the nose to slow down to 70 mph.
3. Push the nose down 30° below the horizon, and, at 100 mph, do a 3-G pull-up.
4. As the nose reaches 5° above the horizon, apply hard rudder in the direction of desired rotation.
5. The instant autorotation commences, let go of the stick and enjoy the ride.
6. When the wings are level again, apply opposite rudder to recover. Having let go of the stick, you can't make the mistake of trying to recover with aileron.

Learning to perform good snap rolls consistently requires a lot of practice, and proper instruction can save the student a great deal of money and time. The best strategy is to learn from a good instructor, practice on your own diligently, and get together with your instructor periodically to refine your technique and nip bad habits in the bud.

You can never really stop learning snap rolls. Seasoned unlimited competition pilots behave like kids in a candy store when they get their hands on a new aircraft type to see how it snaps, and they dissect their experiences as if they were student pilots once again. And many tell you that upon switching to a new competition machine, it took months of practice to learn to snap it to truly competitive standards.

Leo Loudenslager's world-champion snap roll secrets

The best we can hope to do in sharing our experiences on a subject as vast as snap rolls is to lay out the basics and stimulate your imagination to find ways to meet your own challenges based on your personal skill level and the capabilities of the aircraft you fly. Now for some of my own thoughts on this complex topic that none of us can ever perfect but will work endlessly to try to master.

You need to consider several factors as you set out to discover snap rolls. First is the aircraft you fly. There are considerable differences in how well different aircraft perform snap rolls. Some of the important characteristics are the thrust-to-weight ratio, the gyroscopic precession of the propeller (metal vs. composite), wing airfoils, wingspan, the total mass of the aircraft, the mass distribution of the aircraft, elevator and horizontal stabilizer ratios, rudder and vertical stabilizer ratio, and the area of the different control surfaces. The only way to get a sense of where your aircraft fits in is to experiment, take every opportunity to fly different models of aircraft, and compare.

Second, you have to consider the particular type of snap roll you are performing. Control inputs and entry speeds vary depending on whether you are flying a horizontal or vertical snap roll, a snap roll on a descending or ascending 45° line, a ½ or ¾ or some other kind of snap roll.

Third, you have to take into account pilot proficiency and physical condition. The aerobatic pilot is as specialized an athlete as a race-car driver.

Hand-eye coordination and much better physical condition than most people think are both athletes' stock in trade. Our precise control movements and quickness of mind are affected by our physical condition and mental state. The control inputs that worked so well last week or last season might not quite be doing the job on a given day. That might be because they are not precisely the same inputs due to a slightly diminished physical state or lessened ability to concentrate. Don't try to fix the problem by modifying your control inputs. Allow yourself plenty of time to come up to speed again before you change anything.

In my view, speed is probably the single most important consideration for the snap roll. The perfect airplane would be one that is able to maintain a constant airspeed through any type or combination of snap rolls, but that is a utopian dream. In the real world, the higher the power-to-weight ratio, the better off we are, because the closer we can come to snap roll entry speed without having to spend time building extra speed. If we are doing multiple snap rolls, our need for energy is greater, our problem more complex.

1980 World Aerobatic Champion Leo Loudenslager.

Leo Loudenslager

The perfect aerodynamic snap roll is performed at the minimum angle of attack necessary, which gives us a very tight spiral. In fact, it might be so tight that a judge might accuse you of having done an ordinary roll instead of a snap roll. At a world championship, I was flying the Laser and the maneuver was an outside snap roll from inverted to inverted on a climbing 45° line. I hit the maneuver perfectly. Back on the ground, a fellow competitor stated, with some frustration, "That snap roll was either a 10 or a zero." This incident illustrates what you can run into doing the perfect snap roll in competition. The faster roll rates of today's aerobatic machines exacerbate the problem.

With the incorporation of the large leading-edge blunt airfoils first used on the CAP and then incorporated into the Extras, pilots were given what I like to call a larger "sweet spot" for snap roll recovery. Even with this improvement, precise snap roll recovery is a real challenge because of the dizzying rate of rotation. Once you attain a basic proficiency doing snap rolls, perhaps the best single training tool for greatly improving hand-eye coordination is the half snap roll. It is so demanding that if you master it, the rest will be a snap! Good luck and great flying!

Leo Loudenslager is the 1980 World Aerobatic Champion and seven-time U.S. National Champion. He achieved these outstanding results in the Laser 200, an aircraft of his own design and construction in which he continues to fly every kind of snap roll on the airshow circuit, to the crowds' great delight.

4

Vertical rolls

Vertical rolls are rolls performed along a vertical line. The aircraft may be ascending or descending along the vertical line.

The key to flying vertical rolls with precision lies in being able to get the aircraft perfectly aligned on the vertical axis before commencing the roll. This alignment can be best accomplished with the aid of a sighting device mounted on the wing. Some pilots have put a great deal of effort into learning to set vertical lines without a sighting device. After countless hours of practice they are able to do it by reference to the wing and the horizon alone. This accomplishment is admirable, but you will be doing yourself a big favor if you get a sighting device instead.

To get the desired results, the device must be perfectly tuned. Thus, the first task to be accomplished prior to attempting vertical rolls is to check and, if necessary, calibrate the sighting device. This exercise has to be performed with the help of an experienced judge or critiquer.

Generally, the sighting device should be mounted on the side corresponding to the direction in which you perform vertical rolls. Most U.S. pilots prefer to roll to the left and mount the sighting device on the left wing. Some pilots mount sighting devices on both wings.

To establish perfect vertical alignment, you use the bar on the sighting device that is perpendicular to the aircraft's longitudinal axis. When the bar is parallel to and on the horizon, you should be vertical, wings level. If the bar is not parallel to the horizon, the aircraft's longitudinal axis is either positive ("leaning forward") or negative ("leaning aft") of vertical alignment. If the bar is above or below the horizon, the wings are not perfectly level. Let's see how it is done.

1. First calibrate the sighting device to indicate that the aircraft's longitudinal axis is perfectly vertical. Observed by the judge from a distance of about a quarter mile, fly along the X axis of an aerobatic box at 1500 feet. Pull to what your sighting device tells you is vertical (Fig. 4-1) and have the judge critique your performance live, over the radio. Note any required adjustments in the device's alignment parallel to the horizon. To fine-tune any required adjustments, perform a series of vertical lines—up, down (Fig. 4-2), into the wind, and downwind.

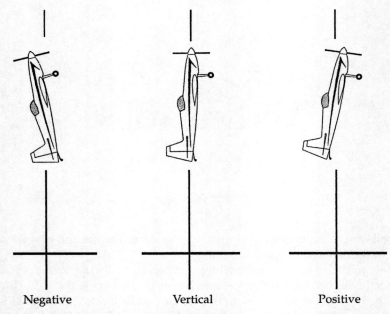

| Negative | Vertical | Positive |

Fig. 4-1. Aligning the sighting device on the X axis, up line. Positive and negative are judge's terms for errors on the vertical line.

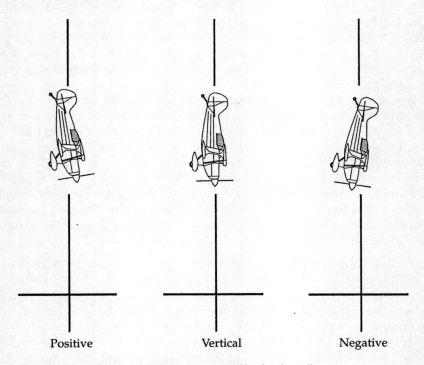

| Positive | Vertical | Negative |

Fig. 4-2. Aligning the sighting device on the X axis, down line.

2. Next calibrate the device to accurately indicate a wings-level condition while on the vertical. Fly along the Y axis at 1500 feet directly at the judge and pull vertical by placing the sighting device's bar squarely on the horizon. Have the judge critique wing alignment to the horizon and note any required adjustments (Fig. 4-3). This adjustment is easily accomplished by sliding the sighting rod up or down along the mounting rod. Some pilots find it easier to establish the appropriate horizon reference by setting the sighting rod slightly above the horizon and marking the horizon itself with tape on the mounting rod.

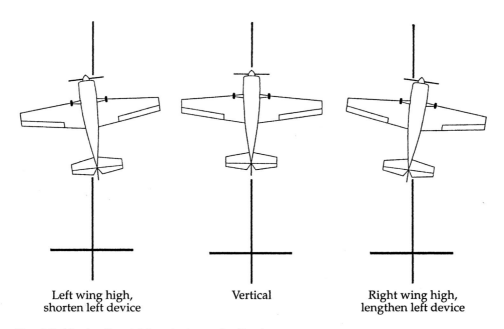

Left wing high,
shorten left device

Vertical

Right wing high,
lengthen left device

Fig. 4-3. Aligning the sighting device on the Y axis.

A word of advice about setting up vertical down lines: They are best entered from a hammerhead. You will have gained a lot of altitude on the up line, and your starting airspeed will be practically zero. Rolling inverted and pulling down to vertical is a bad strategy. The houses get big in the windscreen really fast!

FLYING IT

When you are certain that your sighting device is accurately calibrated, you are ready to begin learning the vertical roll. As we said at the outset, the key to flying precise vertical rolls is being able to establish the perfect vertical alignment of the flight path. Furthermore, this vertical alignment must be established with zero sideslip (we'll say why later). Therefore, the next step is to learn to set a perfect vertical line.

Note that when we say "pull" to vertical, in this discussion we mean pulling to the up line from upright straight-and-level flight. When we say "push" to vertical, we mean pushing to the up line from inverted straight-and-level attitude. While you are learning to set vertical lines, you should never push to the down line from upright straight-and-level or pull to the down line from inverted straight-and-level. The stress on the aircraft and, especially, the speed build-up will be alarming. As we said before, learn to set up the down line from a hammerhead.

Setting the vertical line

At first blush, this sounds like a simple exercise, but is somewhat complicated by gyroscopic precession and slipstream effect. Let's look at pulling to vertical first (Fig. 4-4).

1. From horizontal flight, pitch up into vertical flight with the aggressive application of aft stick. As the horizon disappears under the nose, look along the wing at the sighting device for reference, similar to what you do in a loop. With experience you'll discover that you should actually delay looking at the wing until there is a 70° pitch change to maintain proper nose alignment.

2. As you pitch up, you have to deal with gyroscopic precession. Because of the aggressive pitch change at the high entry speeds required to pull vertical, the propeller's gyroscopic precession causes the nose to move noticeably to the right (on aircraft with right-turning props). Compensate for this by applying slight left rudder until the aircraft is once again in straight-line flight along the vertical axis, at which point no more rudder should be required until slipstream begins to take over as the aircraft slows on the vertical line.
   The amount of rudder required to compensate for gyroscopic precession depends on the entry speed and acceleration (Gs) during the pull or push to vertical. An 8-G pull-up at 170 knots causes greater gyroscopic precession and requires more compensating rudder than an 8-G pull-up at 200 knots, which generates less gyroscopic precession.

3. The rudder requirement in a push to vertical is the mirror opposite of the requirement in the pull. As the nose is pushed, gyroscopic precession causes it to move left (if the propeller is turning right). Therefore, some right rudder is required to compensate.

You should concentrate early in your training on being able to tell if there is zero sideslip on the vertical line. As you become comfortable in the vertical attitude, this task becomes easier. If you need any rudder at all to keep the wings level once established on the vertical, you have sideslip present, which makes it considerably more difficult to perform the roll.

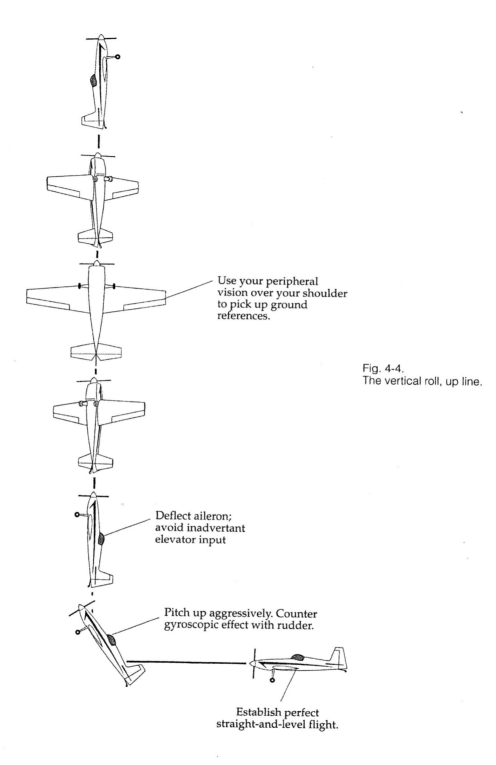

Use your peripheral
vision over your shoulder
to pick up ground
references.

Fig. 4-4.
The vertical roll, up line.

Deflect aileron;
avoid inadvertant
elevator input

Pitch up aggressively. Counter
gyroscopic effect with rudder.

Establish perfect
straight-and-level flight.

Sideslip is caused by two factors on pulling or pushing to vertical:

- Pulling or pushing to vertical with the ailerons deflected
- Not enough or too much rudder to compensate for gyroscopic precession (in some aircraft full rudder might be necessary to compensate)

The roll

Once you have become proficient in establishing vertical flight with zero sideslip, the performance of the roll is a simple process:

4. All you have to do is deflect the aileron in the direction of the roll, taking great care to keep the elevator neutral, and around you'll go.
   No rudder is required when rolling on a vertical line because the angle of attack is zero throughout the roll.
5. In an unlimited competition airplane, the horizon will be a blur in the vertical roll, making it particularly tricky to stop on reference. On the up line, use your peripheral vision to peek down at the ground over your shoulder in addition to concentrating on the wingtip for easier ground reference. On the down line, look straight at the ground over the cowling once you start the roll.

THE FINER POINTS

You can do several things to make learning to do vertical rolls easier. One of the more challenging aspects of performing vertical rolls is maintaining orientation—knowing when to start and finish the roll. To help orient yourself at the beginning of the learning process, set up your first vertical rolls so that you can stop them with the wings parallel to a straight highway, railway, canal or other easily identifiable reference point.

Another technique to further ease the learning process is to perform only half rolls at first. A half roll lasts long enough to show you all the errors and nuances of the vertical roll, yet it does not overload your senses. If you refine the half roll first, you might find it easier to stay fully aware of what the airplane is doing in the full roll.

As you become more proficient, you'll find yourself beginning to rely on internal timing in addition to external references to determine when to start and stop a vertical roll. Strange as it sounds, at the advanced level, aerobatics is a little like dancing. Timing and rhythm become as important as visual reference.

STYLE

For the best visual effect, the vertical roll should have a crisp beginning and an especially crisp, aggressive ending. A well-flown aggressive roll is practically equal to an aggressive stop. This effect is achieved by the aggressive movement of the stick.

To begin the roll, you should smoothly increase the rate at which you deflect the stick. This is subtly but distinctly different from slamming the stick to its stop at a constant rate. If you do the latter, the effect is similar to deploying spoilers, because the instantly deflected ailerons cause drag. If, on the other hand, you smoothly increase the rate of deployment, the ailerons create negligible drag and the roll rate is maximized.

To stop the roll crisply, center the stick as briskly as you can. Drag is not an issue as the ailerons are brought back to neutral, so the centering of the stick at a constant rate is fine.

COMMON ERRORS

Many pitfalls await the novice advanced-aerobatic pilot learning to do vertical rolls, but practice and some good coaching can overcome them all.

Failure to establish proper vertical alignment

There are two common reasons why vertical alignment might not be properly established:

- The sighting gauge is improperly aligned.
- You rush to get vertical.

If you have a sighting-gauge problem (Fig. 4-5), you need to revert to the assistance of a qualified ground observer to determine the degree of error and reset the gauge.

Fig. 4-5
Either the sighting device is off or the right wing is low; you need to know which.

Fig. 4-5. Perfect vertical
alignment. (Continued)

Fig. 4-5. Right wing slightly
high. (Continued)

The rush to getting vertical is easier to fix. Relax! You have more time than you think before you have to start the roll. Take that extra second or two to confirm that you are properly established on the vertical line.

Sideslip

The cause of sideslip can also be found in the pull or push to vertical:

- You use excessive rudder to compensate for gyroscopic effect during the pull or push to vertical.
- You pull or push to vertical with a wing low.

Trial and error will lead you to learning the degree of proper compensation for gyroscopic effect. Pulling or pushing with a wing low is generally caused by looking at the wing too soon during the pitch change. Looking too soon causes you to subconsciously feed in aileron because you have lost the visual clues provided by the movement of the nose. Don't look at the wingtip until you are through 70° of pitch change (Fig. 4-6).

Fig. 4-6
Don't look at the wing when pulling vertical until you reach the 70° mark.

Elevator input on the vertical line

One possible cause of "barreling" the roll might be elevator input on the vertical line. You might be hamfisted on the stick, which is easily checked by reference to the sight gauge. Or, if the gauge appears aligned and you are still barreling the roll, the gauge is misaligned.

Patty Wagstaff on perfecting the vertical roll

Vertical rolls are one of the hardest aerobatic maneuvers to perfect, but they are also the most rewarding. They look spectacular to a crowd of spectators and to the judges when done well and show a lack of practice when not.

When I pull my Extra 300S into the vertical for a vertical roll, I have to be absolutely certain that the wingtips are parallel to the horizon prior to the pull and the airplane is perfectly 90° to the horizon when established on the vertical line. If it is not in the perfect attitude, the rolls will be barreled. Not only will they look sloppy to spectators, but the judges will say, "She wobbled that roll!" and I'll read "Barreled!" on my judging score sheet. And I'll also feel it in the seat of my pants. Wow, that just didn't feel right! I must have put in too much rudder! My hand was pushing the stick forward! My sight gauge must be off!

There are several secrets to doing vertical rolls. Concentration is probably the most important. To keep a marginally unstable aerobatic airplane on a perfect line while rolling around the horizon takes intense focus on a tiny patch of metal touching the sky. Move off that line just a hair and the maneuver is ruined.

Holding the vertical can be easier said than done. The controls of aerobatic aircraft are generally so light and responsive that it takes very little pressure from your hand to make significant applications of the ailerons, rudder, and elevator. A hiccup at the wrong time can cause a large enough unintended control input that the airplane will wiggle or wobble.

It takes intense concentration to keep a light touch on the controls. Pilots tend to translate stress and pressure into their hands and feet, perhaps just because they are trying too hard. A death grip on the stick does nothing to help the cause.

Loosen up. Relax! When I'm feeling stressed, it helps me to open my hand and fly with the palm of my hand rather than the grip of all five fingers. It takes only a light touch to do a good vertical roll and a light one to screw it up.

Advanced aerobatic pilots must know their sight gauge like the back of their hand. They should check it every day before flying to see if it has moved. Sight gauges stick out awkwardly on the ground and are easily bumped into. By checking it every time you fly, you get to know instinctively the angle at which it must be to enable you to do perfect vertical lines.

Lastly, people often ask me how to get a 400° roll rate to stop on a 90° point in a quarter of a second consistently. It takes hours and hours of practice; so much practice that it becomes instinctive. Over time your timing and muscle memory develop to the point where you can do perfect vertical rolls even in conditions so hazy that you can't clearly see the horizon.

Vertical rolls are not only a requirement that advanced aerobatic pilots have to learn to do well consistently, but they are also a joy to perfect. If you can nail your sight gauge on the horizon and keep it there all the way around in the roll, if you can feel no movement in the seat of your pants and can stop on a dime in the sky, you'll reap the ultimate reward, the freedom that only aerobatics can provide.

Patty Wagstaff is three-time U.S. National Champion (1991, 1992, 1993) and a long-standing member of the U.S. National Team. She is also a popular airshow pilot and a qualified movie stunt pilot. She currently flies an Extra 300S.

Patty Wagstaff

5

Hesitation rolls

A hesitation roll is essentially a roll along any line of flight that is momentarily frozen at various points as the aircraft progresses around the roll. Another term for a hesitation roll is, in fact, a point roll. Typical variations are the two-point roll, the four-point roll, and the eight-point roll, all of which can be performed along a horizontal line or up and down vertical and 45° lines. In a two-point roll, the aircraft is paused after rolling through 180°; in a four-point roll, it is paused every 90°, and in an eight-point roll it is paused every 45°.

Hesitation rolls are one of the earliest maneuvers required of aspiring advanced aerobatic pilots as they progress beyond sportsman-level competition. They are relatively easy to perform. They are also showy maneuvers and make a strong impression on judges as they begin to develop a sense for a particular pilot's style. It is therefore very important for the serious competition pilot to learn to perform hesitation rolls well.

FLYING IT

You are not ready to learn to fly hesitation rolls until you are able to perform basic rolls along all lines of flight to a high standard. You need to have an instinctive knowledge of the position of the controls throughout the basic roll as well as during inverted and knife-edge flight.

Momentarily freezing the roll at a particular point to turn it into a hesitation roll is accomplished by momentarily neutralizing the aileron. All other control inputs in effect at the point of hesitation remain unchanged, as does your direction of vision. This sounds easy, but it requires great precision and, hence, a lot of practice (Fig. 5-1). You must be careful not to add up or down elevator as you first neutralize and then reapply aileron. You must also vigilantly maintain visual reference to accurately set the hesitation points.

Spectators and judges alike look for the hesitations to be very crisp in hesitation rolls. When flown by an experienced competitor, the transition between rolling and pausing seems instantaneous. Aspiring aerobatic pilots erroneously

Fig. 5-1
The perfect hesitation roll
from inverted.

Fig. 5-1. (Continued)

Fig. 5-1 (Continued)

believe that this effect is achieved by brutally slamming the ailerons in both directions to initiate the roll as well as to pause it. The technique required, however, is more subtle and complex than that. It should also be already familiar to you, because it is the same aileron technique that is used to start and stop full rolls.

First of all, as in the case of uninterrupted full rolls, a distinction needs to be made between initiating and stopping the aircraft's roll. It is really the stop that has to be crisp for best visual effect. This is, indeed, achieved by slamming the stick to the neutral position as fast as possible, instantly ceasing the asymmetrical lift that causes the wings to roll without creating any drag in the process as the ailerons move to be flush with the wing.

In initiating or reinitiating the roll, however, the situation is different. If the ailerons are brutally slammed to the stops in a constant motion, the initial effect is like deploying spoilers. There is instantaneous drag and strong adverse yaw as the ailerons slam into the smooth airflow over the wings. The drag consumes potential energy (slowing down the airplane), and the strong adverse yaw causes the nose to go off heading. To minimize this effect, the movement of the stick has to be an accelerating motion. As you apply aileron, increase the rate of application. This technique results in a more gradual and smoother transition of the ailerons into the airflow over the wing and markedly reduces drag and adverse yaw. The initial roll rate will be slower, but the increasing rate of aileron application rapidly increases the roll rate.

How can such subtle distinctions possibly be achieved in aircraft with roll rates in excess of 360° per second? With a lot of practice. Advanced aerobatics is like any physical sport played at the highest levels. It takes countless hours of practice to even begin to recognize its subtleties, let alone play the game. By practicing hesitation rolls endlessly, you'll eventually develop muscle memory in your arm and hand that will give you an instinctive "feel" for moving the controls just right.

THE FINER POINTS

You can refine the appearance of your hesitation rolls by ensuring that you spend a sufficient length of time stationary on each point and by developing a feel for the maneuver's rhythm. Without exception, aspiring advanced pilots spend too little time on each point. They hurry through the pauses, which results in the points appearing soft. To enhance a point's crisp appearance, it is important to pause on it for a sufficiently long time to make an impression. As a rule of thumb, always pause on a point for the amount of time it took to get there from the previous point.

When you are doing hesitation rolls on a vertical line (Fig. 5-2), you need to be aware of some nuances to the control inputs and where you should look. On the up line, the aircraft will be slowing; therefore you need slightly more aileron to reach each subsequent point at the same roll rate as you progress through the maneuver. On the down line, the situation is the opposite. The aircraft will be accelerating; the requirement for aileron is less from point to point.

On the up line, it is best to look at the wingtip for visual reference to time the hesitations. On the down line, look straight at the ground. In both cases, as you become more experienced, your sense of rhythm will be a great help in addition to visual references to time the hesitations.

COMMON ERRORS
"Bobbling" the points

Novice advanced aerobatic pilots are inclined to encounter the problem of bobbling the points because they attempt to stop on the points too aggressively to get the crisp effect they are aiming for. The result is that the wing bobbles momentarily on each point. The cause is a deathgrip on the stick and a rigid elbow, which makes the required precision impossible to achieve. Your grip on the stick must be extremely light, your wrist and elbow loose to get the desired results. It's an art, not a craft.

Losing altitude

Losing altitude is caused by being so intent on getting the hesitations right that you forget about the rudder. Remember, you are still doing a slow roll with all the

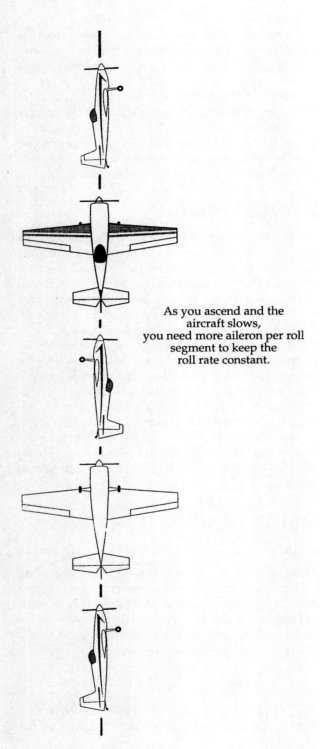

As you ascend and the
aircraft slows,
you need more aileron per roll
segment to keep the
roll rate constant.

Fig. 5-2
The vertical hesitation roll.

elaborate coordination of stick and rudder that it requires. Don't neglect to concentrate on the vision over the nose and the "big picture."

Varying roll rates between points on the vertical line

Varying roll rates between points is caused by the fact that on the up line the aircraft slows, making the ailerons less effective, and on the down line it accelerates, making the ailerons more effective. The only way to cure this problem is to use proportionally more aileron between points on the way up and proportionally less aileron on the way down. "How will I ever learn to tell the difference?" you ask. That's why you need a coach.

Pitts, an American legend

The Pitts Special has done more to further the cause of advanced aerobatics in the United States than just about any other aircraft. It has also won major recognition throughout the world. The story of Curtis Pitts and his feisty little airplane is also a particularly American tale of the rugged individualist on a shoestring budget quietly building a world-class creation in his spare time. The first Pitts Special flew in 1945, but it was not until 1971 that the type finally went into production.

Curtis Pitts, crop duster and one time A&P school operator, loved aerobatics and was convinced that he could build a better aerobatic airplane than the big, underpowered radial-engined machines of the day. The result was the tiny Pitts Special with a 55-hp Lycoming engine. It flew reasonably well, but what really got the Pitts legend going was upgrading its engine to 90 hp shortly after the first few flights. The engine switch transformed a nice little airplane into a rocket ship, and it hasn't looked back since.

Pitts custom-built the first few airplanes himself, refining the design and increasing the engine size to 125 hp along the way. The airplane quickly gained popularity on the airshow circuit in the hands of Betty Skelton, Caro Bayley, and others. Until 1959 the plans for the airplane were largely in Curtis Pitts' head, but he was finally convinced to carefully document the construction of a Pitts by his friend Pat Ledford and offer the plans to the public. This aircraft was designated the S-1C and went on to become one of the most popular homebuilts of the 1960s.

With the advent of the modern world aerobatic championships in 1960, it became evident that vertical penetration and the ability to perform outside maneuvers would be the name of the unlimited game, and Pitts rose to the challenge. During the 1960s, he developed a special set of airfoils that would make the little airplane forget which way was up, added a set of ailerons to the top wings to improve roll rate, and strengthened the airframe to take a 180-hp Lycoming. Known as the roundwing Pitts because of the greater symmetry of its airfoils, this airplane was the S-1S, a world-class unlimited competition machine.

At the 1972 World Aerobatic Championships, held in France, Curtis Pitts' little airplane took the world by storm. Flying Pitts S-1Ss, Charlie Hillard became the men's world aerobatic champion, and Mary Gaffaney took the women's title. The United States also took the team title to make it a clean sweep. For the rest of the decade the Pitts Special was firmly established in the top tier of unlimited world aerobatic competition. When the aircraft of its era were finally forced to yield, among the contenders was another historic American aircraft created in a garage, the Laser 200.

The Pitts S-2B is still a top Unlimited trainer.

Also present at the 1972 Worlds was another Pitts that would play a major role in making Pitts Specials one of the preeminent aircraft types in advanced aerobatics. It was the two-seat, 200-hp S-2A with a constant-speed propeller. It was developed during the late 1960s as a certified production aircraft specifically to offer dual advanced aerobatic instruction on a commercial basis, an activity not permitted in uncertified aircraft. The two-seat Pitts brought advanced aerobatics to hundreds of pilots who would have otherwise only dreamt about it.

The S-2A went into production in Afton, Wyoming, in 1971. The first two customers were Marion Cole and Art Scholl. The S-1S was also certified and joined the S-2A on the production line. It was also offered in kit form over the years. In 1981, the S-1S was upgraded with a 200-hp engine and a constant-speed propeller and became the S-1T. In 1982, the S-2A was superseded by the 260-hp S-2B, and a single-seat version, the S-2S, was also made in limited quantities.

The factory has changed ownership several times over the years, and today it is known as AVIAT. It continues to churn out the popular S-2B and, by special order, the S-2S and the S-1T. AVIAT is also hard at work on the 300-hp S1-11B, destined to challenge the 300-hp monoplanes. And there are still more Pitts Specials flying and competing at all levels in the United States than any other advanced-aerobatic aircraft. The legend continues.

Sportsman competitor Ford Rackemann flying the Pitts S-2B he shares with John Connolly in an aircraft partnership.

6

Outside maneuvers

Outside maneuvers are maneuvers during which the aircraft experiences sustained negative Gs, such as the outside loop and various pushes to and from horizontal and vertical lines. They are in a category of their own among advanced aerobatic maneuvers and are a big step forward in learning to fly advanced aerobatics. In the American aerobatic competition classification system, this step is taken when a pilot transitions from Intermediate to the Advanced category.

Your ability to fly inside maneuvers is no indication of how well you'll do flying outside maneuvers. The biggest challenge in flying outside maneuvers is staying oriented in the strange new perspective they present. They also hurt a lot, especially in the beginning. The high negative Gs exert a force diametrically opposite to the force your body has constantly experienced during your entire life. As they say, "negative Gs turn sport aerobatics into a job," but with practice the discomfort diminishes.

The maneuvers we discuss in this chapter are the outside half loop and the outside loop, which are the best introduction to outside maneuvers. Other outside maneuvers are covered in subsequent relevant chapters: advanced snap rolls, advanced spins, and gyroscopic maneuvers.

FLYING IT

The best way to learn flying outside loops is by starting with the outside half loop. For safety reasons, the outside half loop should be learned by pushing up from inverted straight-and-level attitude rather than pushing over toward the ground from upright straight-and-level attitude. The advantages are self-evident. You are heading away from the ground, and the aircraft slows as it pushes up against the earth's gravity, making it much easier to control speed and avoid overstressing the aircraft if the maneuver goes awry.

The outside half loop from inverted

Here's how the outside half loop from inverted is flown (Fig. 6-1):

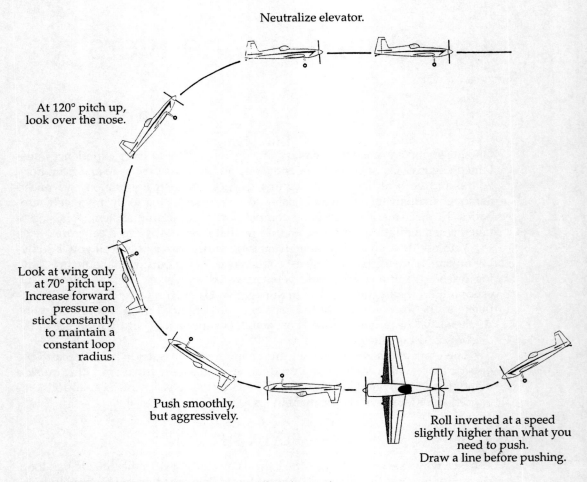

Neutralize elevator.

At 120° pitch up, look over the nose.

Look at wing only at 70° pitch up. Increase forward pressure on stick constantly to maintain a constant loop radius.

Push smoothly, but aggressively.

Roll inverted at a speed slightly higher than what you need to push. Draw a line before pushing.

Fig. 6-1. Start with the outside half loop, pushing from inverted.

1. Begin a dive in upright straight-and-level flight to a speed above the speed you need to push from inverted flight to perform the maneuver. A fair amount of speed is required for the push; make sure you know the appropriate speed for your aircraft. In a high-performance, 300-hp monoplane, around 180 mph is desirable.

2. Half roll to inverted, establish straight-and-level attitude, and set the desired entry speed for the push.

3. To fly the outside half loop, push the stick smoothly, but aggressively, forward. Look over the nose (Fig. 6-2) and use peripheral vision to check that the wings are level. Keep your head back as long as possible, looking at the horizon. The horizon will remain visible up to as much as 70° pitch. As it finally disappears behind you, when you can no longer tilt your head back any further, look at the wingtip to maintain orientation.

Fig. 6-2. In a moment you will switch your vision to over the nose in this outside loop.

The technique for applying forward elevator in the outside half loop is similar to applying aft elevator in an inside half loop. Application has to constantly increase during the maneuver to keep the loop's radius constant. As the aircraft slows in its progress toward the top of the outside half loop, the elevator becomes less effective in the slowing airflow. Therefore, a constantly increasing application of the elevator is required to keep it generating the constant lift required of the horizontal stabilizer to maintain a constant radius in the maneuver.

Initially you need to apply slight right rudder (if the propeller is turning clockwise from the pilot's perspective) to counter gyroscopic precession. If you look at the wingtip too soon, you will find it too disorienting. You will inadvertently feed in aileron in the direction of the wingtip you are looking at (usually left), and without much effort you will go 90° off heading by the time you complete the maneuver.

4. Look at the wingtip until you reach the 120° pitch position; then look over the nose again. Initially you will sense the horizon with your peripheral vision as it reappears on each side of the cockpit; then it will also become visible over the nose.

5. As the aircraft reaches the straight-and-level attitude, neutralize the elevator to complete the half loop. Near the top of the loop the aircraft is slowing down significantly while remaining at a high power setting. As a result, slipstream effect might become noticeable. You might need to aggressively apply substantial right rudder to compensate for slipstream effect.

In the beginning you will find it easier to do a half loop in a monoplane than a biplane because of the better visibility from the monoplane.

The full outside loop

Like the outside half loop, the full outside loop should also be flown from inverted flight initially, for reasons of safety (Fig. 6-3). To fly a full outside loop from inverted, first become proficient in the outside half loop. When you feel comfortable, perform your first full outside loop by flying an inverted half loop and continuing to push down for the full loop. For the first half of the maneuver, refer to the inverted half loop section. To complete the maneuver from the top of the outside half loop, follow these procedures:

1. As you reach the top of the outside half loop, ease off the stick slightly to float over the top, just as you would in an inside loop. You will be slow, which is what you want given the buildup of speed that will occur as you push down.

2. As you float over the top, start pushing over smoothly, constantly increasing forward stick (down elevator) to keep the loop's radius constant. This stick application technique is exactly the same as the application of constantly increasing aft stick in the second half of the upright loop.
   Look straight ahead over the nose at a reference line on the ground to maintain heading.

3. As the aircraft reaches the inverted straight-and-level attitude, briskly neutralize the stick, completing the outside loop. Briefly transition to level flight, and roll rightside up.

As you become proficient with outside half loops and full loops from the inverted position, you can also begin to do them from upright straight-and-level flight (Fig. 6-4), always taking great care to have sufficient altitude and an escape route.

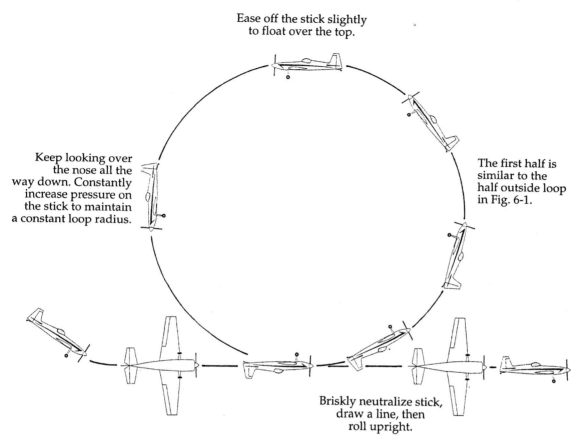

Ease off the stick slightly
to float over the top.

Keep looking over
the nose all the
way down. Constantly
increase pressure on
the stick to maintain
a constant loop radius.

The first half is
similar to the
half outside loop
in Fig. 6-1.

Briskly neutralize stick,
draw a line, then
roll upright.

Fig. 6-3. Progress to the full outside loop pushing from inverted.

THE FINER POINTS

Maintaining wings-level is crucial to accurately flying outside loops. Here is an effective technique to learn how to maintain a wings-level attitude. Roll inverted, stabilize the aircraft in inverted straight-and-level flight, and start a shallow climb while maintaining wings-level by reference to your peripheral vision. It is much easier with this exercise to get used to where the wings should be, instead of trying to sort it all out while attempting to fly outside half loops for the first time.

Outside maneuvers will make you feel lousy in a way the first few sessions of inverted flying might have affected you when you were learning basic aerobatics. As the Gs build, they also hurt. Your vision will get blurred, you will feel pressure in your head and on your eyes, and you might wonder, "Are we having fun yet?" Your body rebels against substantial acceleration from a direction it has never ex-

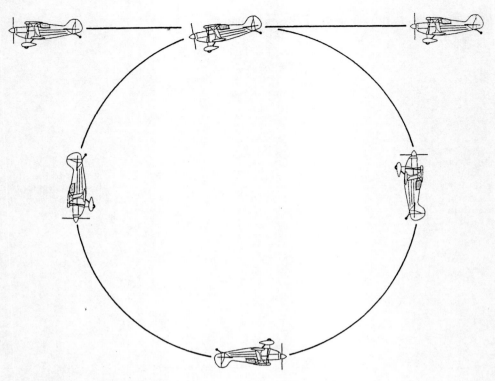

Fig. 6-4. Outside loop pushing from upright.

perienced it before, and it takes some time to adjust. Use common sense; back off if it hurts too much. With practice the effects will diminish and become tolerable, though, quite frankly, for many pilots, they are rarely outright enjoyable.

COMMON ERRORS
Going as much as 90° off heading at the top of the loop

Going off heading at the top of the loop is caused by subconsciously feeding in aileron in the direction of the wing at which you are looking. You are especially prone to committing this error if you have a death grip on the stick that diminishes your ability to feel control subtleties.

The outside loop gets too big

Novices encounter this problem of a loop that is too big because doing the maneuver hurts too much and they don't push sufficiently hard. A big loop diminishes your ability to float over the top and correct for wind, because you end up being too low on energy. The maneuver will end up being pinched.

On the sequence card, this is a dotted line (Patty Wagstaff in her Extra 300S).

Note that we do not suggest that you push harder than what your body can bear. Build tolerance slowly. Work up incrementally to doing the proper-sized outside loop as you build tolerance for negative G.

The bottom of the loop is hooked

The cause of a hooked loop is pushing excessively at the bottom of the maneuver because of a fear of building up too much speed. Again, don't be unsafe. If you are hooking the bottom of your outside loops, be cautious in easing off or pushing too hard. Maintain careful speed control. Work with a coach.

7

Rolling circles

Rightly or wrongly, the Germans have a reputation for concocting complicated things. It should come as no surprise that the rolling circle, possibly the most difficult of all aerobatic maneuvers, was invented by Gerhard Fieseler, a German pilot in search of a new challenge in the 1930s. He is the innovative aerobatic pilot who also developed the first successful inverted system.

A rolling circle (also known in aerobatic jargon as a "roller") is a maneuver in which the aircraft is rolled while flown in a turn of 90° or more. In the maneuver's most basic form, the aircraft is rolled four times while completing a 360° turn. Each roll takes up one 90° segment of the turn. Thus, if the aircraft continuously rolls into the turn from straight-and-level attitude, the roll pattern throughout the turn is as follows: level at the beginning; inverted at 45°; level at 90°; inverted at 135°; level at 180°; inverted at 225°; level at 270°; inverted at 315°; and level at 360°, back where it started.

Don't let this talk of degrees confuse you. It is best to think of rolling circles in segments. In the above example, the aircraft has to complete a full roll on each 90° segment of the 360° turn. If you look at it that way, it is easy to grasp that at the beginning of a 90° segment the aircraft is level, at the halfway point of the segment (45°) it is inverted, and at the end of the segment (90°) it is once again level. This sequence is repeated three more times to complete the maneuver.

The upright inside rolling circle with four rolls is only one example of this maneuver. Three or two rolls on the circle are other variations (the fewer the rolls, the harder the maneuver, as you will see). The rolling circle can also be commenced from inverted, and the rolls can be flown to the outside of the turn.

The objective of the rolling circle is to keep the roll rate, the turn rate, and the altitude constant while hitting the cardinal points of the maneuver in the right attitude. This complex requirement demands a superb sense for timing the movement of aileron, rudder, and elevator. Don't be frustrated if initially you seem to be getting nowhere. Just to be able to fly a barely passable rolling circle requires hours and hours of diligent practice.

FLYING IT

The rolling circle has a certain rhythm to it that makes it easier to accelerate into the maneuver from a moderate airspeed rather than decelerate into it. Maneuvers that have slow exit speeds, such as a half loop, are ideal for setting up a roller. The best way to learn the rolling circle is to learn one segment of the various permutations of the four-roll rolling circle. Accordingly, learn to fly one roll on a 90° segment the following ways:

- From upright, rolling inside (into) the turn
- From inverted, rolling inside the turn
- From upright, rolling outside (away from) the turn
- From inverted, rolling outside the turn

The roller is a fluid maneuver, requiring a continuous manipulation of all three controls in a rhythmic pattern. "Like stirring a pot of soup," say some of its more experienced practitioners. Keep this in mind as you read the next section.

A 90° inside rolling circle to the left from upright

1. Establish straight-and-level flight and look out to the left from the cockpit. Pick an easily recognizable landmark at the 90° point (off the wingtip) and the 45° point. As the aircraft progresses through the maneuver, it should be inverted when the nose is pointing at the 45° mark and upright again when the nose reaches the 90° mark (Fig. 7-1).

2. Having selected your landmarks, look directly at the 45° point. Apply left rudder to get the nose moving in the turn. As you yaw to the left, the right wing wants to come up naturally; aid it with a little left aileron. The aircraft will be in a slight skid, but it will be both rolling and turning.

3. As the aircraft approaches knife-edge, apply slight aft elevator to continue the turn (like in any roll, the elevator is now acting as the rudder). On knife-edge you should be halfway to the 45° landmark (keep looking at it).

4. As the aircraft approaches inverted, you have to switch your field of vision from the left to the right side of the cockpit to continue looking at the 45° landmark. You now need to add right rudder to keep the aircraft moving, turning in the right direction (if you left in the left rudder, it would reorient the turn away from the desired direction). Apply forward elevator to keep the nose on the horizon. If you timed all this right, you will be perfectly inverted as the nose reaches the 45°, landmark, and you will be at the altitude where you were when you started the maneuver.

5. "So far, so good," said the man as he passed the 34th floor after he jumped off the Empire State Building. But now you are inverted, and you still have 45° of turn to go.
   Actually, at least on paper, the remainder of the maneuver is quite simple. Except for the aileron input, which stays constant, just do in reverse everything you did to get here.

Continue looking out the right side of the aircraft and shift your focus to what was the 90° mark when you started the maneuver; with only 45° of turn remaining, is now really the new 45° mark, but we will continue to refer to it as the 90° mark. Continue applying right rudder to keep the turn going. Maintain left aileron to keep the roll going.

6. As the aircraft approaches knife-edge, apply increasing forward elevator to keep pushing the nose toward the 90° landmark and left (top) rudder to keep the nose from dropping.

7. As the aircraft approaches upright flight, completing the 90° segment of the rolling circle, you need to make a decision. If you want to keep going with another 90° segment, stay on the rudder and, without losing the fluid rhythm, get ready to repeat the cycle. Upon reaching the 90° landmark, you should have neutral elevator, left rudder, and the slight amount of left aileron that you had throughout the roll; in other words, you'll be perfectly set up to continue the rolling circle.

If you want to exit the maneuver at this stage (as you would while learning its basics), ease off left rudder, ease off forward elevator, and resume straight-and-level attitude. We'll stop right here, and so should you until you become more proficient. But just think: Three more rolls at this stage, and you'll have completed your first four-roll 360° rolling circle! Good luck!

A 90° inside rolling circle to the left from inverted

This maneuver is no different from the one described above. Rather it is a matter of choosing to divide the full 360° rolling circle into four quarters by starting at a point at which the aircraft is inverted. Thus it follows that to fly this maneuver, you simply have to fly the inverted portion of the roll first, followed by the upright portion, and continue until you finish up inverted (Fig. 7-2).

Therefore, refer to the above section for flying the maneuver, but start with the maneuver from step 5, followed by steps 6, 7, 1, 2, 3, and 4.

There is an advantage to starting the rolling circle from inverted that appeals to many competition pilots. The maneuver commences and finishes inverted, affording a splendid view of the ground below and making positioning in the aerobatic box much easier than from upright flight.

A 90° upright outside rolling circle to the left

To fly an outside rolling circle to the left, you need left rudder to start the turn, aft elevator to keep the nose from dropping, and right aileron to start the roll. Apply forward stick as the aircraft starts to roll and approaches knife-edge. Increase left (top) rudder to keep the nose on the horizon.

At this stage you are doing the maneuver on internal timing, because you rolled away from your 45° landmark and can't see it. Pick up the reference as soon as you can as the aircraft rolls past knife-edge (Fig. 7-3).

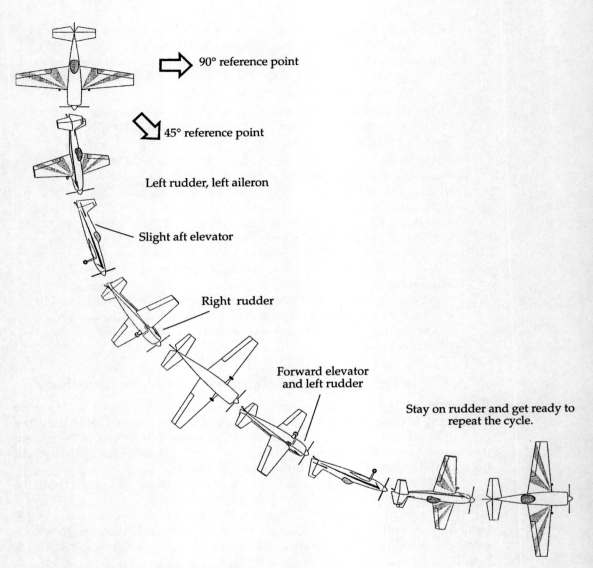

90° reference point

45° reference point

Left rudder, left aileron

Slight aft elevator

Right rudder

Forward elevator
and left rudder

Stay on rudder and get ready to
repeat the cycle.

Fig. 7-1. A rolling circle to the inside from upright.

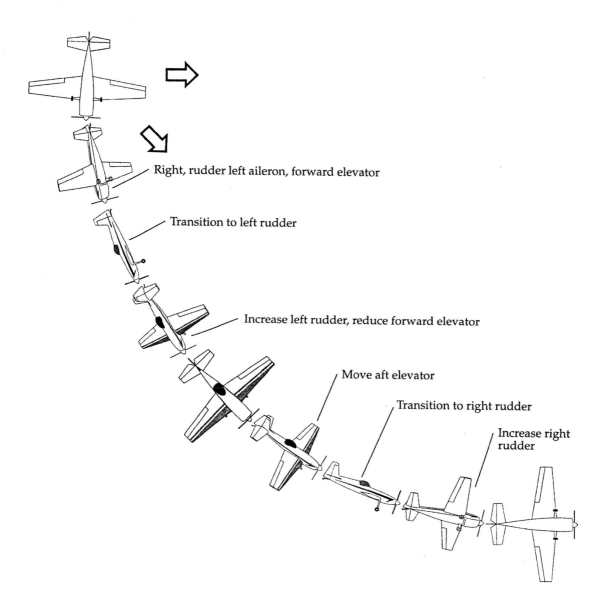

Right, rudder left aileron, forward elevator

Transition to left rudder

Increase left rudder, reduce forward elevator

Move aft elevator

Transition to right rudder

Increase right rudder

Fig. 7-2. Rolling circle to the inside from inverted.

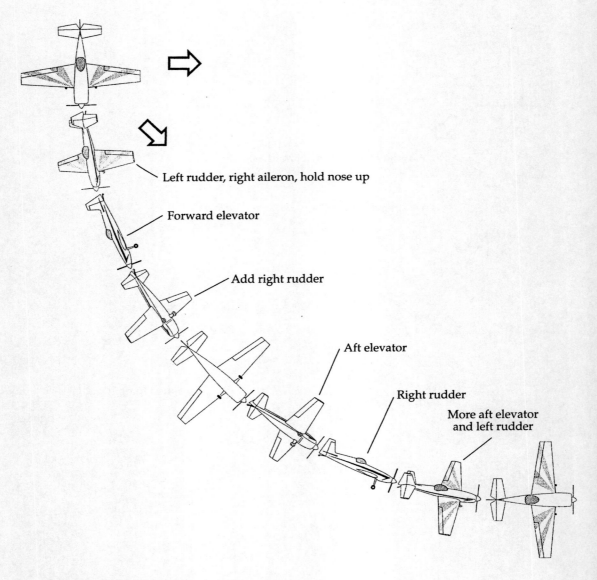

Left rudder, right aileron, hold nose up

Forward elevator

Add right rudder

Aft elevator

Right rudder

More aft elevator
and left rudder

Fig. 7-3. Rolling circle to the outside from upright.

From the inverted position, you need to transition to right rudder and aft elevator to complete the maneuver. Elevator and rudder applied simultaneously toward the inside of the circle speeds up the roll rate considerably. To keep the roll rate constant, you have to compensate by decreasing aileron input proportionally.

A 90° outside rolling circle to the left from inverted

This maneuver is similar to the upright outside maneuver; only the sequence is different. Start with the inverted portion of the upright outside rolling circle first, and finish with its upright portion (Fig. 7-4).

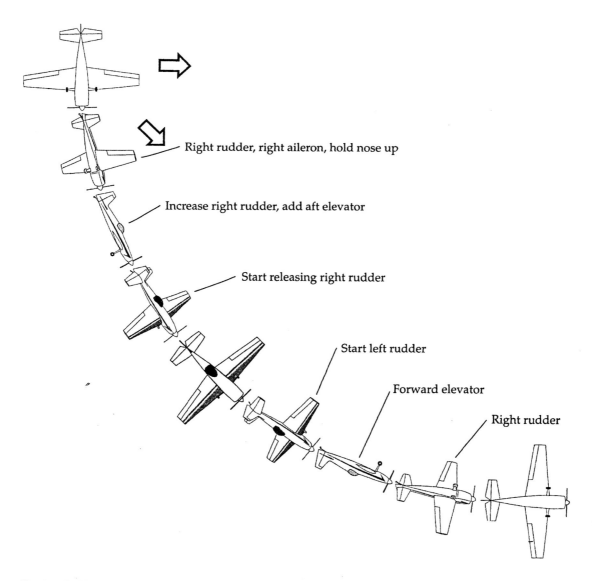

Right rudder, right aileron, hold nose up

Increase right rudder, add aft elevator

Start releasing right rudder

Start left rudder

Forward elevator

Right rudder

Fig. 7-4. Rolling circle to the outside from inverted.

THE FINER POINTS

Take your time learning this maneuver and concentrate on picking up its rhythm. Just as you first slowly move through the steps of a new dance, move the controls slowly as you begin to explore the rolling circle. This results in a very wide turning radius, and you might end up abandoning a given attempt, but it is a good way to concentrate on and absorb the required control movements and their consequences. Slowly increase the pace of control movements as you begin to see what is going on, and eventually you'll find yourself picking up the natural rhythm of control movements required to fly the maneuver.

There are two variations of the four-roll 360° rolling circle that appear in competition: the three-roll and two-roll rollers. The former requires a roll every 120°, the latter every 180°, with midpoint landmarks at 60° and 90°, respectively. These maneuvers might sound easier than four-roll rollers, but in fact they are not. Timing the larger segments is more challenging.

The two-roll roller especially requires a lot of patience. The aircraft slows down considerably during the maneuver, requiring much more elevator and rudder input. Significantly decaying airspeed makes the second 180° particularly hard to fly well.

COMMON ERRORS

The most prevalent error in performing rollers is the failure to use a sufficiently harmonious blend of rudder and elevator to accomplish a smooth, continuous turn. In the examples, all rolling circles are flown to the left.

The airplane snaps out of the maneuver

If excessive rudder or elevator is used during any roller, you'll notice because the airplane will simply snap out of the maneuver. If you keep ending up in inadvertent snap rolls, blend the use of rudder and elevator more evenly.

The maneuver looks segmented

The roller looks segmented if you use too much elevator. With excessive elevator and not enough rudder, you'll be turning only on knife-edge. Some pilots deliberately grind their way through this maneuver in this fashion and take the judging consequences, finding it preferable to consistently perform a passable, if mediocre, maneuver instead of risking a zero score.

The roll rate accelerates

The roll rate accelerates on those segments of the maneuver where rudder and elevator are both applied toward the inside of the circle. For example, while rolling inside from upright to inverted, the rudder and elevator are both applied inside, (left rudder, aft elevator), and the elevator aids the rudder's effect, accelerating the roll during the final stage of this segment of the maneuver. To compensate, you need to ease off the aileron a touch.

When rolling outside, the nose drops (dishes)

You'll encounter the nose dropping at the beginning of the maneuver when rolling to the outside from upright to inverted. You need left rudder to start the turn, followed by forward stick (down elevator) to keep it going. But forward stick causes the nose to pitch down. The way to fix this problem is to slightly raise the nose prior to commencing the maneuver.

8

Tailslides

In a tailslide, the aircraft is flown on a vertical up line and is allowed to decelerate until it comes to a stop and slides back about one fuselage length. It then swings around its lateral axis ("swaps ends"), is established on a vertical down line, and is pulled or pushed into the next maneuver.

There are two types of tailslides: Stick forward, also known as wheels-up, and stick back, also known as wheels-down. In the stick-forward tailslide, the aircraft pitches aft in the pivot. In the stick-back tailslide, it pitches forward. This might initially strike you as contradictory. See if you can figure out why it is not a contradiction before you read the answer in the following section.

For safety reasons, we consider the tailslide exclusively an Unlimited maneuver. In the U.S. competition system, it is found only in Unlimited programs. This might sound disconcerting, and indeed, if a tailslide is flown with precision, it is a perfectly safe maneuver. However, if the aircraft is allowed to slide back too far without swinging through its lateral axis, and it picks up excessive speed, damage can occur in the tail area. In many aircraft, the elevator support and hinge points were not designed to withstand the forces generated by high-speed reverse airflow. It is easy for even experienced competition pilots to get into a situation in which there is the potential for damage; hence the Unlimited classification.

The tailslide (like all the other maneuvers in this book), should never be self-taught under any circumstances. Become proficient in performing tailslides with a competent advanced aerobatic instructor before flying them solo.

FLYING IT

Stick-forward and stick-back tailslides differ only in the initiation of the pivot at the top of the maneuver (Fig. 8-1). We therefore first discuss the stick-forward tailslide and then present those aspects of the stick-back tailslide that are different.

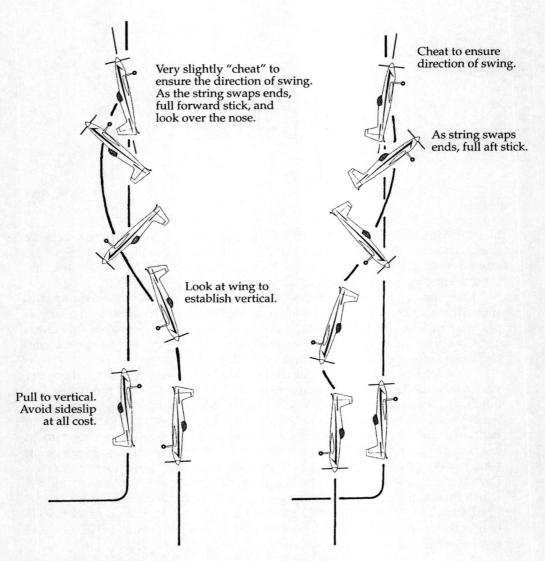

Very slightly "cheat" to ensure the direction of swing. As the string swaps ends, full forward stick, and look over the nose.

Cheat to ensure direction of swing.

As string swaps ends, full aft stick.

Look at wing to establish vertical.

Pull to vertical. Avoid sideslip at all cost.

Fig. 8-1. The tailslide: stick-forward (left) and stick-back (right).

1. You can enter a tailslide from positive (upright) or negative (inverted) flight. Pull or push to a vertical up line. The minimum airspeed required for a tailslide is equivalent to the hammerhead entry speed of the same aircraft. Be exceptionally careful to have zero sideslip. This maneuver is extremely sensitive to sideslip. The slightest loading on the up line will cause a wing drop in the slide. Refer to Chapter 4 for a refresher on establishing vertical lines with zero sideslip.

2. You looked out at the wingtip to establish the vertical up line. Continue doing so, and reduce power to idle to let the aircraft slow. Watch the string on the sighting device (Fig. 8-2).

Fig. 8-2.
Note the yarn on the sighting device, used to indicate airflow reversal in the tailslide.

3. When the perfectly vertical aircraft is almost stopped, apply the slightest touch of aft stick. All you are looking for is the most minute hint of a change in attitude. Were you to measure it, it would amount to no more than 3°. This control input sets up the aircraft for the swing aft. It is necessary because if the aircraft is allowed to remain perfectly vertical, you can never know with certainty which way it will swing, regardless of your stick input to induce the swing. In essence this amounts to what pilots

refer to as the "cheat." In the strictest interpretation of the rules, it is irregular, but if you don't do it you'll regularly run the risk of zeroing the maneuver by swinging the wrong way.

4. The aircraft will come to a stop and start sliding downward along the vertical line. The string on the sighting device will swap ends as the slide begins. Wait for an instant to allow the aircraft to slide, then gently and slightly ease the stick forward. This control input causes the aircraft to pivot aft (wheels up). As you apply forward stick, shift your vision to straight ahead, over the nose. The reason forward stick causes the nose to pitch aft instead of forward is that the airflow over the horizontal stabilizer and elevator is reversed, resulting in control reversal (Fig 8-3).

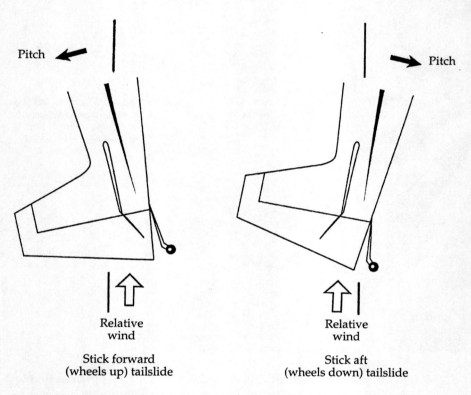

Fig. 8-3. The effects of airflow reversal over the horizontal stabilizer.

5. The aircraft will now pivot aft, pendulum through the vertical, and end up with its nose pointed down, more or less vertical. Look at the wingtip and precisely establish the vertical down line. Let the aircraft draw the down line long enough to build up the proper speed to transition to the next maneuver.

To perform a stick-back (wheels-down) tailslide, the only control inputs that have to be done differently are the ones to set up and initiate the pivot:

1. An instant before the aircraft stops on the up line, "cheat" with very slight forward stick to set up the direction of the pivot.

2. Allow the aircraft to slide for an instant as the string swaps ends, and then apply slight back stick. The aircraft will pivot nose down, pendulum through the vertical, and settle more or less along the vertical down line. Look at the wing to establish perfect vertical, allow the aircraft to draw the line, and pull or push into the next maneuver.

THE FINER POINTS

When flying with an aircraft equipped with a metal propeller, you should reduce power earlier on the up line than when flying with a composite or wood propeller. The reason is that the metal prop is heavier. It has more inertia and can cause torque effect if the rpms are not sufficiently reduced. The earlier the power is reduced, the more time the prop has to slow down while the aircraft ascends on the up line and comes to a stop. Excessive torque can cause the fuselage to rotate around the prop at the top of the tailslide when the aircraft is practically stopped (hanging by the prop), wreaking havoc with alignment in the maneuver. An aircraft with a light airframe and a heavy metal propeller, such as the Pitts, is understandably the most prone to torque effect in the tailslide.

Aircraft with wood or composite propellers are not immune to torque effect in the tailslide. However, their lighter propellers slow down more quickly than metal propellers because of their lower inertia. At any rate, it is always best to reduce power on the up line as soon as possible.

When you are planning your freestyle sequences, you can save yourself a lot of difficulty by not placing snap rolls and rolling circles directly before a tailslide. At the end of these maneuvers, the aircraft might retain some side load, making it practically impossible to pull up into the tailslide with zero sideslip. Good planners of compulsory programs are also aware of this issue and take it into account when designing the sequences.

Some pilots prefer to "cheat" along the entire up line to set up the direction of the swing. They are either minutely positive or negative all the way up. We feel that this technique is more noticeable to the judges and, therefore, prefer the minor adjustment at the top instead.

COMMON ERRORS
Failure to reduce power soon enough

If power is reduced too late on the up line, the propeller won't have a chance to fully slow down, and some residual torque effect will remain, causing the aircraft to twist around the longitudinal axis as it begins to slide. This problem is more of an issue for aircraft with heavy metal propellers and the short-coupled Pitts.

Sideslip

As is the case with vertical rolls, it is crucial to have zero sideslip on the up line. The slightest sideslip will cause the aircraft to drop a wing in the slide.

Sideslip is introduced on a pull or push that is too aggressive, creating so much gyroscopic effect that it becomes difficult to totally eliminate it with anything less than perfect handling. Enter the tailslide into the wind, which enables you to corner with a radius of sufficiently small appearance with less Gs. You will find the lower level of gyroscopic effect easier to manage. If you do end up with sideslip, correct with inboard rudder and aileron.

Wing low on the vertical line

Having a wing low is a problem that is prevalent on hazy days. When there is a lot of haze, the horizon can appear to be lower than it is. If you place the sighting device where you think the horizon is, you might end up with one wing low. What you need to do is look straight ahead until 70° pitch-up to maintain nose alignment, and then look at both wings to properly set the sighting devices on the true horizon.

Know your parachute

The parachute is a rarely used but vital piece of emergency equipment. While some segments of aerobatic sequences are flown too close to the ground for the parachute to make a difference, more often than not there is ample time and altitude in an emergency to use the parachute. Many an aerobatic pilot has floated to earth to tell the tale and fly again. Most of us know little about our parachute beyond how to strap it on, where the rip cord is, and when to get it repacked. We might think that all security chutes are created equal, but there is a lot more to it than that. To maximize your chances on the day that we hope will never come, it is worthwhile to consider how to select the right parachute and how to look after it.

Like airplanes, parachutes have structural operating limits. One of the most important limits is the maximum speed at which a parachute can open without being damaged. For security chutes, it is best to get one rated at 150 knots or higher, because airspeed can build rapidly under conditions that require emergency egress.

Once you are satisfied that your chute can withstand the blast as it opens, the next important consideration is how fast it will descend with your weight suspended under it. Too fast a rate of descent increases the chances of broken limbs or worse on touchdown. The rate of descent for a given weight depends on the inflated diameter of the parachute (dimensions are not always given in inflated terms, but the information is available if you request it). The greater the inflated canopy, the slower the descent. And the heavier you are for a given parachute, the faster you'll come down. According to conventional wisdom, the desirable descent rate is 16 feet per second (ft/sec) or less. A descent rate of 16 ft/sec is 960 ft/min, more than a "firm arrival" in an airplane, so you can see why it is important to get it right for your weight when you are dangling at the end of a parachute. Consult the performance information for your weight carefully before you make your selection.

It is advisable to get a parachute that is steerable and know which lines are the steering lines (they are usually a different color). When you buy the parachute, have an experienced rigger inspect it for you, just as you'd have your mechanic check out an airplane before you bought it.

Once you have your parachute, you can take several simple steps to preserve its condition. Keep it in a dry dark place when not in use, away from ultraviolet and infrared light. Keeping the chute in the cockpit inside a hangar is fine only if your hangar has very low humidity. Hang up your parachute by the straps in the open air for a few hours once a month to air out any moisture that might have accumulated.

Include the parachute in your preflight. Check if the pins are properly installed and make sure that none of them are bent. Have your rigger show you what to look for on your particular brand of parachute. And from time to time practice on the ground how you'd get out of the cockpit if you ever had to.

9

Advanced snap rolls

A number of variations of the basic upright snap roll are staple maneuvers in advanced aerobatic competition. They are generally similar in execution to the basic snap roll, but each has its own set of special demands that sets it apart and requires a concentrated effort to master. In this chapter we address the vertical snap roll, outside snap rolls, and fractions of snaps.

VERTICAL SNAP ROLL

The vertical snap roll is, as the name implies, a snap roll along a vertical line. Fundamentally, it is similar to a horizontal snap roll with some subtle differences in execution. Two main differences set the vertical snap roll apart from the basic horizontal snap roll.

The first difference is that the aircraft is much more prone to losing energy in the vertical snap roll; therefore it is exceptionally important not to "bury" it in the stall. The arc inscribed by the nose during the maneuver has to be as tight as possible to minimize energy loss, requiring a more aggressive acceleration of the maneuver. Any error has greater negative consequences than in the horizontal snap.

The second difference is that the pilot's visual orientation is very different in the vertical snap roll. In all other snap rolls, the pilot looks directly over the nose for visual clues. In the vertical snap roll, there is nothing to see over the nose, so the pilot has to look at the wingtip and aft, at the ground, to maintain orientation.

Because of the much greater energy loss on the up line, the entry speed for the vertical snap roll must be considerably higher than for the basic horizontal snap. In the highest-performance 300-hp-plus monoplanes, the minimum entry speed is 180 mph; 200 mph is preferable.

On the vertical line, the two segments of the line before and after the snap must be of equal length. The tendency is to snap too soon, which traps the pilot into having to draw a very long line after the snap. It is also important to take your time to carefully establish the aircraft on the vertical, because a stabilized aircraft rewards you with a cleaner snap.

The seat-of-the-pants feeling that the pilot experiences in the inside vertical snap roll is less than it is in the basic upright horizontal snap roll, but it is there and is a valuable aid to developing a feel for the maneuver. You should, therefore, learn to do full snaps before fractions of snaps to evolve your sense of timing.

Timing becomes especially important in marginal visibility when clues on the horizon are obscured. In poor visibility, you have to combine a sense of timing with a focus over your shoulder on the ground. As you gain experience, it is not unusual in a well-marked box to fly vertical snaps by relying almost entirely on the panel markers.

In good visibility, the view beyond the sighting device gives you your primary visual clues, and it is particularly important to get the "big picture." Don't focus on the sighting device. Look through it at the horizon. Try to develop an awareness of the entire space around the airplane, and imagine the airplane's proper alignment in it.

FLYING IT

When you learn to fly vertical snap rolls, start by doing full snaps, and do them first on the vertical up line (Fig. 9-1). Full snaps have a seat-of-the-pants feel to them, due to the autorotation, with which you should become familiar before proceeding to fractions of snaps. The minimum entry speed required to do one full vertical snap roll on the up line is 180 mph in either an advanced aerobatic monoplane or a Pitts.

1. Set up the aircraft in straight-and-level flight perpendicular to a highway or similar reference line. Establish your entry speed so that you can pull vertical just as you reach the highway.

2. Pull to vertical just as you would for a vertical roll, maintaining zero sideslip. Look at the wingtip for reference.

3. Sharply bring the stick back and, a split second later, aggressively apply full rudder in the direction of the roll to get the airplane into an accelerated stall and initiate autorotation.

4. As autorotation commences, simultaneously add forward stick and slight aileron in the direction of roll to accelerate the snap roll. To maximize acceleration and thereby minimize planform drag by reducing the radius inscribed by the nose around the roll, you need more forward stick than in the horizontal snap roll.
   Aileron input should be slightly less than in the horizontal snap roll because the aircraft is slowing on the up line and you want to minimize any drag caused by the aileron deployment as much as possible.

5. Hold full rudder until the instant recovery is desired. At this point, forcefully apply opposite rudder until the autorotation stops. Neutralize

rudder and add more forward stick to recover. Additional forward stick is required to reestablish the aircraft on the vertical line. During the snap, it had to remain off vertical throughout the maneuver to maintain the accelerated stall condition.

The vertical snap roll on the down line is performed in exactly the same way as it is on the up line, with the exception that you should look straight over the nose at the ground for reference (Fig. 9-2).

THE FINER POINTS

Don't worry if you initially feel that you are not making much progress with developing your sense of internal timing. You will find, as with horizontal snap rolls, that it is only after having done hundreds of snap rolls that you acquire an internal sense of timing for the maneuver, which is caused by the consistency of the duration of the autorotation. Once this sense of timing is developed, you no longer need to rely too much on outside references to fly the maneuver. At this point, you will also find it much easier to time ½ snaps and ¾ snaps.

You can employ another neat trick when you are doing a vertical snap roll into a stiff wind. You can actually get away with staying slightly positive (about 2°). The wind pushing you backward creates an illusion to the judges that you are perfectly vertical when, in fact, you are slightly positive. The advantage of being slightly positive is that in relation to the vertical line you need to pitch up to a lesser extent to reach the critical AOA (you need less additional G since the wing is already slightly loaded). You will have less planform drag and will therefore lose less energy. And when you are flying balls to the wall, every bit of energy makes a difference.

COMMON ERRORS
The snap roll is performed too soon

Performing the snap roll too soon is the most frequent mistake made by novice advanced aerobatic pilots. It seems that they are afraid of running out of energy if they wait too long, so they rush the maneuver. Remember, if your entry speed is correct, you will have plenty of energy to perform the snap in its proper place after having drawn a suitably long vertical line.

Failure to accelerate the snap roll sufficiently

In the vertical snap roll, there is a much lighter sensation of G. The feeling is that practically nothing holds you into your seat, so the tendency is to try and hold on to the stick, which subconsciously makes it difficult to push it away from you with the authority and for the distance required to properly accelerate the snap.

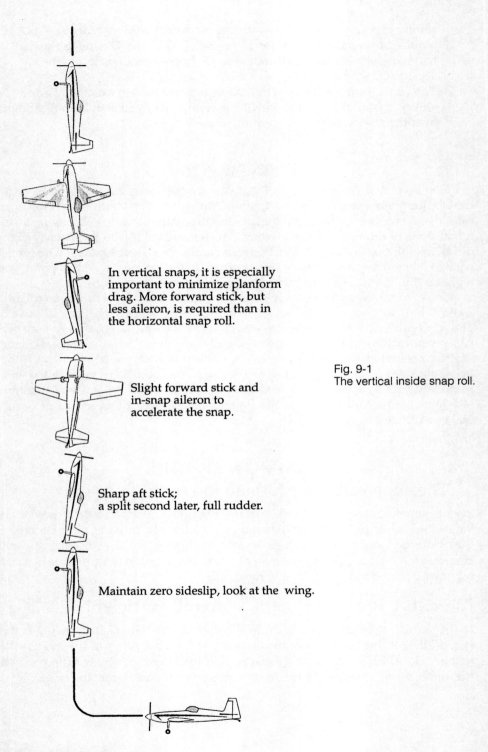

In vertical snaps, it is especially important to minimize planform drag. More forward stick, but less aileron, is required than in the horizontal snap roll.

Fig. 9-1
The vertical inside snap roll.

Slight forward stick and in-snap aileron to accelerate the snap.

Sharp aft stick; a split second later, full rudder.

Maintain zero sideslip, look at the wing.

Fig. 9-2. It is easier to find references for the snap roll on the down line than on the up line.

Inability to perform the maneuver without straight reference lines on the ground

Pilots who learn the maneuver in a place that has generous straight reference lines on the ground often are unable to perform the maneuver when they show up at a contest where that is not the case. To forestall this problem, first learn the maneuver parallel to a reference line, and then practice it perpendicular to the line, transitioning to a hammerhead at the top to see how well you did stopping on heading.

OUTSIDE SNAP ROLLS

Outside snap rolls hurt, perhaps more than any other maneuver. This fact causes pilots to approach them with some concern, especially if they are generally uncomfortable in inverted flight. But if you are highly proficient at flying inverted, then outside snaps won't seem all that different from inside snaps. Don't attempt to learn outside snaps until you are as at home inverted as you are rightside up.

Horizontal outside snap roll

When you are learning outside snaps, start with the horizontal outside snap roll (Fig. 9-3), and set up the maneuver from a half loop with a layout to inverted. Look over the nose throughout the maneuver for orientation.

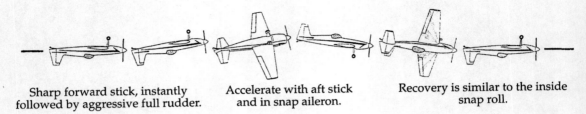

Sharp forward stick, instantly followed by aggressive full rudder.

Accelerate with aft stick and in snap aileron.

Recovery is similar to the inside snap roll.

Fig. 9-3. The horizontal outside snap roll.

Initiate the snap from about 115 mph in a top unlimited monoplane. Because you are inverted, the direction of stick movement and aileron movement is reversed in comparison to the corresponding inside snap, and less aileron is required, but the control movement sequence is the same.

1. Once established in straight-and-level inverted flight at the correct entry speed, sharply bring the stick forward to raise the nose just beyond critical angle of attack (AOA). It is not necessary to bring the stick full forward to accomplish this condition. How far forward the stick comes is determined by your airspeed and how aggressive you are on the elevator. The faster you are, the less forward stick you need (to counter the wing's lift vector), however, the faster the movement of the stick needs to be to break the relative wind. As in the corresponding inside snap, you have to be careful not to go too far beyond critical AOA. If you do exceed critical AOA by a big margin, you'll be buried in the stall.

2. As you reach about 50 percent of the pitch-up above the horizon toward critical AOA, aggressively apply full rudder in the desired direction of rotation. The idea is to have full rudder deflection as critical AOA is achieved. At this instant, autorotation commences.

3. Once the aircraft starts to autorotate (a split second after you apply full rudder), accelerate the snap by performing the following control inputs simultaneously:

   • Add aft stick to the point at which the angle of attack just exceeds the critical angle of attack (the aircraft is just on the edge of flying but is not).

   • Add an extremely slight touch of aileron in the direction of rotation. Remember, the aileron input has to be opposite to the rudder input because you are inverted.

The margin for error in using aileron is very slight, much less than in the corresponding inside snap. Excessive aileron will result in a spiral instead of autorotation.

4. Hold full rudder until the instant recovery is desired, which is when the aircraft reaches the attitude it had upon commencing the snap roll. At this point, forcefully apply opposite rudder until the rotation stops (you initiated autorotation with rudder; therefore you have to stop it with rudder). Neutralize rudder and stick, and you'll be flying straight-and-level inverted and on heading.

Vertical outside snap roll

The vertical outside snap roll compares to the horizontal outside snap roll in the same way the respective horizontal and vertical inside snaps relate to each other. The control inputs are the same in the vertical outside snap as they are in the horizontal outside snap. The vertical outside snap also has to be accelerated more aggressively than the horizontal outside snap because of the greater loss of energy on the vertical up line. The entry speed for the vertical outside snap is the same as it is for the vertical inside snap.

The downside of the vertical outside snap is that it really hurts. The upside is that there is greater visibility, making it easier to maintain orientation.

COMMON ERRORS
The stall is not initiated aggressively enough

Because the outside snap hurts so much, many pilots wince at aggressively setting off the accelerated stall. The fact is, however, that the more aggressively you set off the stall, the quicker the Gs are over and the better off you are. It is like taking medicine. Get it over quickly.

The outside snap roll is not accelerated sufficiently

Insufficient acceleration occurs because the pilot, hanging by the straps under the effect of negative G, subconsciously wants to hold on to something, which turns out to be the stick. Relax the death grip, and make a conscious effort to accelerate the proper amount.

FRACTIONS OF SNAPS

Common fractions of snaps are ½ and ¾ snap rolls. They are tricky maneuvers, primarily because if the autorotation is stopped before completing a full revolution, the aircraft generally ends up off heading.

Take, for example, the ½ snap. For some reason that is not entirely clear, most pilots have a bias toward snapping to the right. But, because of gyroscopic precession, as the nose is pitched up, it moves to the right. Therefore, if we snap to the right, the nose goes further off heading, and if we stop the snap at the halfway point, we'll be about 15° off heading. We can cure this problem in one of two ways:

- We can start the snap by placing the nose 15° off heading in the direction opposite the snap. When we stop the snap halfway through, we'll be on the desired heading.
- We can correct with momentary left rudder immediately after we initiate the snap with full right rudder. However, we have to reapply right rudder so that we stop the snap aggressively enough to get the desired crisp effect.

There is a much easier way to cure the problem of going off heading in the right ½ snap roll: Do the maneuver to the left! In this case, gyroscopic precession is your ally. As you pitch up, the nose goes right due to gyroscopic precession. As you yaw left to perform the autorotation, the movement of the nose to the left cancels out the movement to the right caused by gyroscopic precession. When you stop the snap at the halfway point, the aircraft ends up on the correct heading.

Another essential technique to help prevent the aircraft from finishing off its heading is to accelerate the ½ snap much more and at a much more rapid rate than the full snap. This technique tightens the radius of rotation and reduces the heading change.

Now consider the ¾ snap roll. It is hard to precisely stop on heading, and if all you do is stop it with rudder, the aircraft ends up with the in-snap wing low. The fix is to ease off the snap rudder slightly as the aircraft rotates past the ½ snap point. But you have to be careful, because reduced rudder authority yields a "soft" finish if you stop the snap as you stop a snap in which you have full-rudder authority. To achieve the desired brick-wall effect, you have to stop the snap with an extra hard application of rudder.

These techniques are equally applicable to vertical and outside fractions of snaps. With the 360° to 450° roll rates that are so common today, you'll have a lot of fun trying to figure out how to do accurate fractions of snaps. It is for this reason that world champion Leo Loudenslager considers the ½ snap roll the best training maneuver for learning to snap.

COMMON ERRORS

The recovery is bobbled

The problem of a bobbled recovery is another one involving the death grip. The pilot is too tense, holds the stick too hard, and presses the rudder pedal too rigidly. The tension is caused by the need to be alert because of the lightning speed at which the maneuver happens.

Acceleration is insufficient

Insufficient acceleration is generally the result of not reacting quickly enough to the rate at which the maneuver is taking place. By the time the pilot gets around to accelerating the maneuver, it is over and the aircraft is buried. Get that stick forward the instant autorotation commences.

From a garage to the top of the world: Story of a world-class monoplane

Today the factory-built Extra 300S is one of the best and most well-established unlimited aerobatic aircraft in world-class competition. Yet when Walter Extra set out to design and build his first unlimited aerobatic monoplane, it wasn't with the backing of a major aircraft manufacturer or deep-pocketed investors. He had only his own knowledge, resources, and energy to rely on, and the inspiration of three Americans before him who, in the tradition of Curtis Pitts, had built their own unlimited aerobatic monoplanes in their garages in their spare time because the market wasn't offering an airplane that met their needs.

The trail leads back to the late 1960s and the Stephens Akro, designed by Clayton Stephens for U.S. women's champion Margaret Ritchie. The idea was to gain an edge in performance over the Pitts by building an equally small and light midwing monoplane that had a lot less drag and a faster roll rate. One Akro builder was Leo Loudenslager, who completed his airplane in three years, finishing it in 1971.

The Akro served Leo well as he steadily built competition experience at the Unlimited U.S. nationals until 1975, when he decided to give the airplane a higher performance wing with a new airfoil. He took the opportunity to also substantially modify the fuselage. Only about 5 percent of the Akro remained when Loudenslager was done. The net result was an essentially new airplane, which he named the Laser 200. It was much lighter than the Akro and had a stronger spar and a larger rudder, among many other differences. Nearly 50 percent of its weight was the engine and the propeller. It had the power-to-weight ratio of the Pitts and had a lot less drag. Its roll rate was phenomenal, aided by "shovels" on the ailerons, another Loudenslager innovation that became popular over the years. The redesign proved an instant hit. Loudenslager won the 1975 U.S. National Championships in the Laser 200 and six more in the next seven years! And in 1980 he claimed the ultimate aerobatic prize when he became the world aerobatic champion.

Leo Loudenslager's legendary Laser 200.

The Akro was also highly modified by another top American competition pilot, Henry Haigh, a contemporary of Loudenslager. Haigh retained the original airfoil but modified the wing structure and the tail structure. He called his airplane the Superstar. At the 1982 World Championships, Haigh flew the Superstar to second place, missing first by only 5 points out of 16,404, and in 1988 he became the world champion in it.

Inspired by the Laser's and the Superstar's design and already an experienced Pitts builder, Walter Extra set out to build a monoplane of his own creation, the 230-hp Extra 230, with a wing design derived from airfoil research done by German aeronautical engineers during World War II. Furthermore, he would custom-build the Extra 230 for anyone who wanted one. Making its debut in 1984, the Extra 230 outperformed the competing monoplanes of its day, and a steady stream of orders led Walter Extra from being a homebuilder to becoming a leading manufacturer of advanced aerobatic aircraft. Today when you look at an immaculate Extra 300S, it is hard to believe that it all started with a couple of determined homebuilders in the proverbial garage.

10

Advanced spins

We addressed basic upright spins in detail in our first aerobatic book, *Basic Aerobatics*. We dealt with flying spins for spin training, performing upright spins in competition, and handling inadvertent spins that might surprise you during the course of maneuvering. We devoted a lot of space to emergency spin-recovery techniques, chief among them the Beggs/Müller emergency spin recovery. While we present a brief summary of the basic upright spin here, it would be most useful if you also took a moment to review the spin chapter in *Basic Aerobatics* before tackling advanced spins.

Also bear in mind that different aircraft might recover differently from various spins. Always follow and become highly proficient in the manufacturer's spin recovery procedures presented in the aircraft operating manual before trying other techniques, such as the Beggs/Müller emergency recovery technique. If your aircraft operating manual leaves you in any doubt about recovery from any form of spin, consult the manufacturer.

Spins are as an important part of advanced aerobatics as they are of basic aerobatics, often seen in competition flying. The aspiring advanced aerobatic pilot must become proficient in several advanced forms of spinning. The most common advanced spin is the inverted spin. Inverted spins are a staple of all advanced aerobatic competition events. There are also the accelerated spin and the flat spin, in both upright and inverted modes. These spins are seen only in the four-minute freestyle, because they are impossible to precisely stop on heading consistently, which makes them unsuitable for conventional competition. In addition to intentionally flying all these spins with confidence, the novice must also learn to recognize them and recover from them if they are entered inadvertently.

Before we proceed to advanced spins, let's briefly review the basic upright spin. The spin is an autorotational maneuver. To induce the spin, you must first slow down the aircraft by bringing the throttle to idle. Then you pitch up the aircraft to exceed the critical angle of attack. As the stick approaches full aft, you simultaneously yaw with rudder in the desired direction of rotation.

The yaw causes the inner wing to slow down even more and the outer wing to accelerate. The lift generated by the slowing inner wing diminishes in relation to the lift generated by the accelerating outer wing. (Both wings are beyond critical

AOA; nevertheless, they continue to generate some lift, but not enough to continue "flying." The inside wing drops as the aircraft rolls along the longitudinal axis (Fig. 10-1). The inside wing's AOA increases further while the outside wing's AOA decreases because of the shift in relative wind induced by the roll. The inside wing generates more drag, which is added to the drag caused by the rudder-induced yaw. The asymmetric drag causes the aircraft to rotate around its vertical axis. The autorotation along the two axes continues as long as the control inputs are maintained. At this point the controls are full aft stick, full rudder in the direction of yaw. The throttle is at idle. The aircraft is in a spin. The aircraft's simultaneous movement around the longitudinal and vertical axis yields a resultant axis that is known as the spin axis (Fig. 10-2).

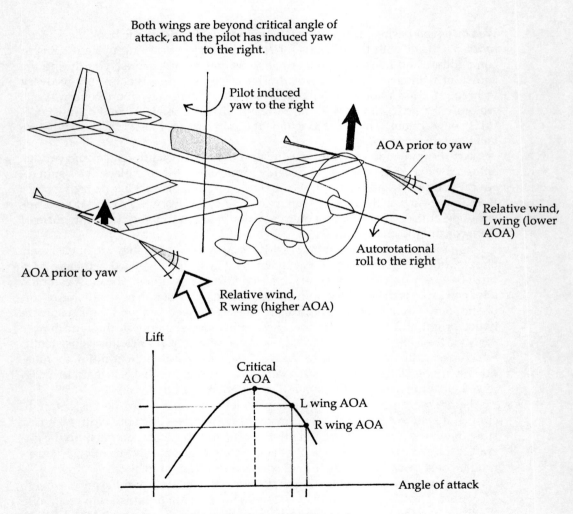

Fig. 10-1. Wing drop and the commencement of autorotational roll.

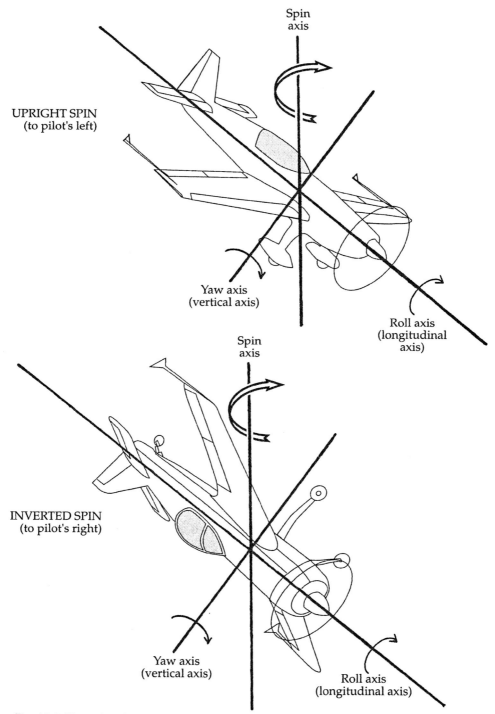

UPRIGHT SPIN
(to pilot's left)

Spin axis

Yaw axis
(vertical axis)

Roll axis
(longitudinal axis)

Spin axis

INVERTED SPIN
(to pilot's right)

Yaw axis
(vertical axis)

Roll axis
(longitudinal axis)

Fig. 10-2. The spin axis.

To recover from the spin, you must stop the yaw (to equalize drag on the two wings and stop rotation), and you must break critical AOA. Here is the recovery sequence (confirm that the power is at idle):

1. Apply full rudder opposite the yaw and wait for the rotation to stop.

2. When the rotation stops (and only then), apply forward stick (down elevator, forward of neutral) to break critical AOA.

3. At this point, the aircraft will be flying again in a dive, accelerating rapidly. Pull out of the dive, but be careful not to pull so hard as to induce an accelerated stall and possibly an inadvertent secondary spin.

Before we move on to advanced spins, let's discuss two important points in more detail than we did in *Basic Aerobatics*.

Why is it important to keep aft stick (up elevator) in the spin until the rotation stops?

It is most important in aircraft with the conventional horizontal/vertical stabilizer layout that is found on most aerobatic aircraft (as opposed to T-tail layout), for one simple reason (Figs. 10-3 and 10-4). To stop the yaw most effectively, you need maximum available rudder. If the stick is pushed forward, applying down elevator, the effective rudder surface exposed to the relative wind in the spin is reduced. Therefore, if down elevator is applied before the yaw stops, less effective rudder surface is available to stop the yaw. It will at best delay stopping the yaw and at worst make it impossible.

Note that aircraft with low-mounted stabilators can be notoriously difficult to recover from a spin because aft stabilator almost completely blankets the lower part of the rudder.

What really happens aerodynamically when forward stick (down elevator) is applied to break critical AOA in a developed spin?

When down elevator is applied, it is pushed down against the relative wind. The relative wind pushes back up against it, reestablishing a new equilibrium. In the process the nose is swung downward, and critical AOA is broken (Fig. 10-5).

INVERTED SPINS

The aerodynamic forces acting on an aircraft in an inverted spin are exactly the same as they are in an upright spin. The only difference is that the aircraft is inverted, requiring different control inputs to induce the spin. A simpler example of the same principle is inverted straight-and-level flight. Thrust and drag, lift, and gravity are exactly the same as in upright flight, except they now act on the inverted airframe.

The inverted spin is not more difficult to fly than the upright spin, but the different perspective requires some getting used to by the aspiring advanced aerobatic pilot. Let's see how it is flown.

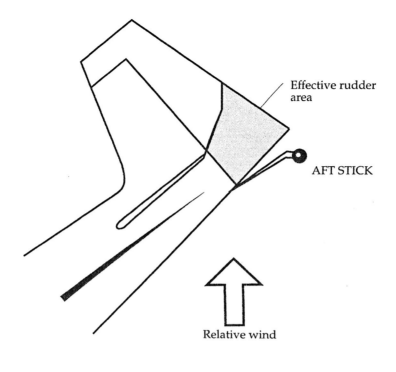

Effective rudder area

AFT STICK

Relative wind

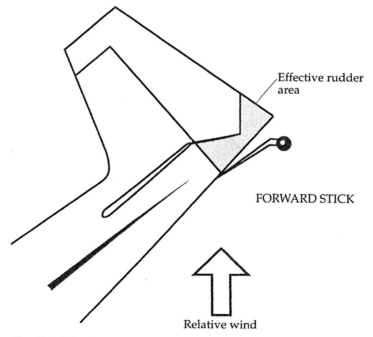

Effective rudder area

FORWARD STICK

Relative wind

Fig. 10-3. Why the rudder is more effective with aft stick in an upright spin.

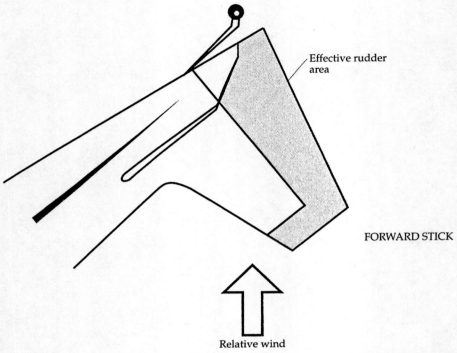

Effective rudder
area

FORWARD STICK

Relative wind

Fig. 10-4. Why the rudder is more
effective with forward stick in the
inverted spin.

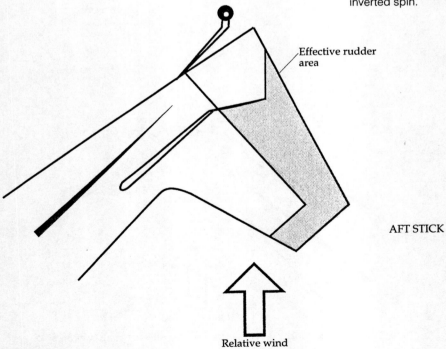

Effective rudder
area

AFT STICK

Relative wind

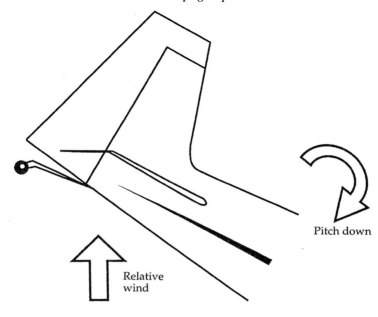

Forward stick in upright spin

Pitch down

Relative wind

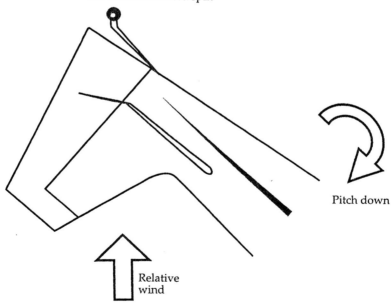

Aft stick in inverted spin

Pitch down

Relative wind

Fig. 10-5. The relative wind striking the changed elevator causes the nose to pitch down during spin recovery.

Flying it

The inverted spin is best set up from a preceding maneuver that leaves the aircraft at a slow airspeed, such as a layout off of a vertical line or from a half loop to inverted. The 1½-turn inverted spin is also the most common inverted spin flown in competition.

1. Establish straight-and-level inverted flight at the top of a vertical maneuver, after a half loop, directly over a good reference line such as a straight highway. While not difficult to fly, the inverted spin can be quite disorienting to the novice, so a solid, clearly identifiable reference line is very important to aid recovery. Retard the throttle to idle. Look over the nose (continue looking over the nose throughout the maneuver).

2. Ease the nose up with down elevator (forward stick). Continue to add forward stick until you feel a buffet as the aircraft approaches the critical angle of attack. Maintain altitude by reference to the altimeter. It is easier than trying to do it by reference to the horizon as the nose pitches up. This is a rare occasion in aerobatics when you fly by reference to the altimeter, instead of just casting an occasional glance at it. Now the aircraft is ready to spin. Note that the stick is not yet fully forward.

3. As the aircraft begins to buffet just prior to the stall, add full rudder in the desired direction of rotation. As the inside wing drops, apply full forward stick (down elevator) to keep the aircraft stalled and have maximum rudder authority.
   As autorotation sets in, the aircraft will stabilize in a fully developed inverted spin.

To recover, you have to accomplish exactly the same things as you do in the upright spin:

- You must stop the yaw to equalize drag on both wings.
- You must break critical AOA.

These tasks are accomplished as follows:

1. Apply full opposite rudder to stop the yaw.

2. As the rotation slows significantly, smoothly apply aft stick (up elevator) to break the critical AOA.

3. As the rotation stops, neutralize all controls. The aircraft will now be flying again and will be more or less vertical.

4. Establish a precise vertical down line and pull or push out of the maneuver.

In most advanced aerobatic aircraft, it will take anywhere from ¼ to ½ turn to recover from the inverted spin.

The finer points

One of the first things you'll notice about inverted spins is that the rotation is more rapid than in upright spins. This is because there is more effective rudder surface available to maintain autorotation (see Fig. 10-4). For the same reason, recovery from an inverted spin is also more rapid than from an upright spin.

As you discuss inverted spins with your peers, you might get into heated discussions about whether or not yaw and rotation are opposite each other in an inverted spin. The reason for contrary opinions is one of perspective. A critiquer who is rightside up on the ground observing the aircraft in an inverted spin will say the aircraft is spinning to the left (the critiquer's left). Yet this so-called spin to the "left" was induced by the pilot (who, along with the aircraft, is inverted) with right rudder. So was the aircraft yawed to its "right" to rotate to the critiquer's "left"? The critiquer might see it that way, but as the pilot, you must look at the world from the pilot's perspective. So just remember this:

- If you apply left rudder, the aircraft will always rotate to your left. From your perspective, you are in a left spin, although the critiquer, from his or her perspective on the ground, might call it a spin to the right.
- If you apply right rudder, the aircraft will always rotate to your right. From your perspective, you are in a right spin, although the critiquer on the ground might call it a spin to the left.

Note that this interpretation differs from what we said in *Basic Aerobatics*, where we defined left and right from the critiquer's perspective. We have since come to believe that the best way to clear up this confusing matter is to always talk from the pilot's perspective. If you get into a heated discussion about this topic, the first question to ask to sort it out quickly is "whose perspective?"

ACCELERATED AND FLAT SPINS

You can perform accelerated spins as well as flat spins independently of each other, but as you will see, there is a natural progression from the plain spin to the accelerated spin to the flat spin. We therefore believe that in learning to perform accelerated and flat spins it is most effective to first learn how to accelerate a spin and then learn how to make that accelerated spin go flat.

Due to gyroscopic precession caused by the propeller's rotation, spins can be made to go flat only opposite the direction of the propeller's rotation, and they will also accelerate more effectively in that direction (they can, however, be accelerated in either direction). Therefore, we'll assume that you are flying an aircraft with a right-turning propeller (the majority of modern aerobatic aircraft) and will discuss accelerated and flat spins accordingly:

- Upright accelerated and flat spins to the pilot's left
- Inverted accelerated and flat spins to the pilot's right

If you fly a Sukhoi, Yak, or any other aircraft with a left-turning propeller, you need to substitute left for right and vice versa for control inputs throughout this discussion.

Upright accelerated and flat spins

In an accelerated spin, the aircraft usually has its wings almost level, its rate of rotation is much higher than in a normal spin, and its nose is slightly further up toward the horizon than in a normal spin. In a flat spin, the nose is pitched up even further, to an attitude practically level with the horizon.

As we discussed earlier, pilot-induced pitch, roll, and yaw are all essential tools for making an aircraft spin. We use the controls and power (which includes gyroscopic effect) to change the magnitude of pitch, roll, and yaw developed in the basic spin to turn it into accelerated and flat spins.

In a developed spin, the rate of autorotation is stabilized. The "spin" and "antispin" forces are in equilibrium. Increasing the "spin" forces in a basic spin to accelerate autorotation is the objective of the accelerated spin. Rotation stabilizes at a higher rate when the "spin" force and the "antispin" force reach a new state of equilibrium.

The unequal lift components of the inside and outside wings in the basic spin add the roll component to the spin, which is stabilized at a certain roll rate. We can induce a change in this state of equilibrium by applying out-spin (right) aileron. Out-spin aileron reduces the difference between the lift component on the two wings by reducing the difference in AOA (which remains above critical for both wings; consider the relative wind), and reduces the roll component. However, the down left aileron causes greater drag than the up right aileron. This imbalance accelerates autorotation.

Autorotation is now hindered by the drag caused by the lower surface of the rudder (the section below the elevator) exposed to the relative wind, which is now from below as the spin has become stable. Down elevator (forward stick) greatly reduces the rudder surface below the elevator, significantly reducing the rudder's drag. Exposed to greatly diminished "anti-spin" forces, the autorotation accelerates further. The aircraft is in an accelerated spin.

As the autorotation accelerates, the nose pitches up slightly, due to the increased centrifugal force. Advanced aerobatic instructor Randy Gagne has a great method of demonstrating this effect. Tie a tennis ball to a long string at the end of a stick. Hold the stick vertical and slowly rotate the tennis ball. It will find a level of equilibrium as it rotates at a low constant speed. Now increase the speed of rotation. What happens to the tennis ball? It rises. This is exactly what happens to the aircraft's pitch attitude as the spin accelerates.

To turn the accelerated spin into a flat spin, the objective is to raise the nose even further. To do so, simply apply full power. The propeller's gyroscopic precession brings the nose further up.

The nose of a short-coupled airplane, such as the Pitts, will not pitch up as high as the nose of a long-coupled airplane, such as the Extra 300S, as should be obvious to anyone who understands weight-and-balance calculations. But enough of all this discourse. Let's go flying!

Flying them

1. To fly an upright accelerated spin, first put the aircraft into a fully developed basic upright spin to the left. Look over the nose and continue doing so during the entire maneuver. If you try to look at the blur of the horizon for a landmark, you'll just feel ill. When the spin is fully developed, you are ready to accelerate it.

   Apply full out-spin (right) aileron. Doing so has two important consequences:

   - Roll is significantly reduced because the lift component on the two wings is almost equalized.
   - As roll is reduced, yaw is simultaneously increased due to the greater drag of the left (down) aileron. The spin begins to accelerate as the autorotation forces become stronger, and the wings will become almost level.

2. Smoothly apply forward stick (down elevator). The reduction in the rudder surface below the elevator exposed to the relative wind diminishes rudder drag, further accelerating the spin.

   The nose also comes up due to the increased centrifugal force (the "tennis ball on a stick" effect). Applying forward stick and seeing the nose come up might feel disconcerting at first, but that's the fun of advanced aerobatics. Straight-and-level "rules" don't always apply.

   When the rotation stabilizes at a higher rate, you are in a fully developed accelerated spin.

If you are flying a Pitts, you should try something that might surprise you. (Don't you love surprises in the middle of an accelerated spin?) Apply full right rudder. NOTHING HAPPENS! There is so much rudder area in the dead air above the elevator that the rudder is now completely ineffective. (If you try it in a monoplane, the rudder retains some effect.)

Now let's turn the accelerated spin into a flat spin. It's easy.

3. Simply apply full power and watch the nose come up almost level with the horizon (more in a long-coupled aircraft than in one that is short-coupled). Note that the rotation rate is not affected because the propeller's gyroscopic precession acts parallel to the vertical axis.

You can also do a flat spin without first doing an accelerated spin by not applying forward stick. Let the basic spin develop, and apply out-spin (right) aileron and full power. The spin goes smartly flat, but the rotation rate is slower than in the accelerated flat spin, and the nose does not come up as high.

Let's return to the accelerated flat spin and see how to recover from it. Once you are used to the very high rates of rotation, it is rather easy. The technique is based on the Beggs/Müller emergency recovery technique:

1. Reduce power to idle. If you didn't add power to the accelerated spin to make it go flat, the power should already be at idle.

2. Let go of the stick. COMPLETELY! This is extremely important! The stick will come aft and centered as the relative wind pushes the elevator up and the differential drag caused by aileron input is equalized. The rudder will resume its role as the primary source of yaw and the aircraft will return into a conventional basic spin.

3. Apply opposite rudder, and when the rotation stops, apply forward stick to break the critical AOA and pull out of the dive.

Inverted accelerated and flat spins

What happens to the aircraft in inverted accelerated and flat spins is no different from what happens to it in the upright version of these maneuvers. However, because the aircraft is inverted, different control inputs are required to achieve the same effect. Keeping in mind the differences in control inputs between conventional inverted and upright spins, read the section below. You should find it quite easy to follow. We first accelerate the conventional inverted spin, and then we make it go flat.

Flying them

1. Start by placing the aircraft into a conventional inverted spin to your right (right rudder).

2. As the inverted spin is developing, you are ready to accelerate it. Apply in-spin (right) aileron to reduce the difference in the lift component between the wings and greatly reduce the roll. Asymmetric drag caused by the aileron deflection (right aileron down, left aileron up) causes the autorotation to continue. The rotation will begin to accelerate.

3. Now, to minimize the drag caused by the rudder, you need to reduce the rudder area under the top surface of the elevator (remember, you are inverted) exposed to the relative wind. Accomplish this by smoothly moving the stick aft (up elevator). The spin accelerates further. The nose pitches up slightly because of increased centrifugal force (the "tennis ball effect" again). When the rotation stabilizes at a faster rate, you are in a fully developed accelerated spin.

4. To make the accelerated spin go flat, add full power. The nose rises approximately to the horizon, depending on type of aircraft.

If you want to go into an unaccelerated inverted flat spin, skip the aft stick input. From a conventional inverted flat spin, apply in-spin (right) aileron and full power. The nose won't pitch up as much, and the rotation will be slightly slower.

Recovery is similar to the upright inverted flat spin, accomplished by the Beggs/Müller emergency recovery technique:

1. Reduce power to idle. If you are only in an accelerated but not a flat spin, the power should already be at idle.

2. Let go of the stick COMPLETELY. As in the upright version of the maneuver, this is very important. The stick will center itself and move forward, the nose will drop, and the aircraft will reestablish itself in a conventional inverted spin. Rudder will again be the primary source of yaw.

3. When the nose has dropped, apply full opposite (left) rudder. When the rotation stops, apply aft elevator to break the critical AOA and pull out of the dive.

Advanced spinning is very challenging, but if you plan to do advanced aerobatics, particularly advanced outside maneuvers and gyroscopic maneuvers, it is mandatory from a safety standpoint that you become fully proficient in all kinds of advanced spins, inadvertent or intentional. It is guaranteed that as you learn and practice advanced maneuvers, you'll repeatedly find yourself in inadvertent advanced spins of all sorts, and when that happens you had better know exactly what to do. Every time.

COMMON ERRORS

The inverted spin is attempted before the pilot is ready

If you fly the inverted spin before you are totally comfortable flying inverted, you'll scare yourself, you'll find it difficult to learn the maneuver, and you'll get depressed by the lack of your progress. As is the case with outside snaps, be as at home inverted as you are upright before tackling inverted spins.

Failure to develop internal timing for performing inverted spins

The inverted spin is more disorienting than the upright spin. You can't see the horizon well and have few if any good reference points. To perform inverted spins to competition standards accurately, you therefore have to develop a good sense of internal timing for when to stop the maneuver.

The accelerated spin is accelerated before it is fully developed

This problem occurs when you try to rush the accelerated spin and try to accelerate the spin before it is fully developed. The result of this error is simply that the aircraft will recover from the spin instead of transitioning into the accelerated phase of the spin.

Failure to apply the required amount of control inputs in a flat spin

Failing to apply the required amount of control inputs in a flat spin can be dangerous. It is caused by the excessive centrifugal force that presses you into your seat. You have a tendency to subconsciously apply the controls with the same feeling of force that you are accustomed to when not subjected to strong centrifugal force. The result is that you might think that you are applying sufficient force to fully move a control, when in fact you are not, and the control is only partially applied.

As a result of this phenomenon, you might fail to retard the power fully to idle and might not apply full opposite rudder on recovery, even though you are convinced that you are doing everything right. You need to consciously verify that the throttle is full aft and the rudder is applied to the stop. If you don't, the aircraft might not recover, and you might end up jumping out of a perfectly good airplane, which is embarrassing, to say the least.

In some aircraft, depending on cockpit layout and the length of your leg, you might end up not having enough leverage to apply full rudder against the centrifugal force. To prevent this situation from occurring, always make sure prior to takeoff, when you are strapped in, that your leg presses down on the rudder in the neutral position at a 30° angle relative to the airplane's longitudinal axis. This angle ensures that you have enough leverage.

Bob Meyer in the family Phoenix biplane.

Advanced Aerobatic Maneuvers in History

Advanced aerobatic maneuvers came into being for a variety of reasons, most having to do with self preservation. Less than a decade after the Wright brothers' first successful flights, the most daring pilots were feeling confident enough to experiment with exploring the structural limits of their machines to find out how far they could safely go. Records are often sketchy, and some maneuvers were flown inadvertently, without being well understood. Thus the question of who was first to fly a particular maneuver is not always clear. It is certain, however, that most classic advanced aerobatic maneuvers were flown regularly by the end of the 1920's.

Understandably, the first advanced maneuvers to be discovered in the woefully underpowered early machines were ones that relied on gravity for energy, such as the tail slide. Maneuvers requiring strong vertical penetration, such as vertical snap rolls, were the last to be conquered. The air battles of World War I were a powerful influence on the development of advanced maneuvers, especially the basic snap roll and spins. Maneuvers useful primarily for show, such as the hesitation roll, emerged only when competitive and display aerobatics became the rage. Let's take a brief look at the origins of some of the most common advanced maneuvers.

Tail Slides

Adolphe Pegoud, the great French aerobatic pioneer was the first to conduct a deliberate series of flight tests (for Bleriot) exploring the tail slide in 1913. A year before Pegoud's tests, Englishman WBR Moorhouse also stumbled into the beginnings of a tailslide when he decided to see what would happen if he held his Bleriot vertical and just sat there. The tail's vigorous rattling and shaking scared the early pilots, and many opted for the safer escape of the wingover before the tailslide had a chance to develop. However, by 1914, the tailslide was a standard aerobatic maneuver.

Snap Rolls

Snap rolls were born of the need to escape a stream of bullets coming at you from behind in the unfriendly skies of World War I. Stick back and into a corner, full rudder in the same direction, and the snap roll was born as the Red Baron cursed "Vere did the schweinhund go?!" The first one was surely a desperate reflex reaction.

Spins

Spins were first entered into inadvertently (Parke's Dive) and were greatly feared. A few brave souls, chief among them Sopwith's Australian test pilot Harry Hawker, took it upon themselves to conquer this killer. They were the first to discover that usually all would be well if they cut the power, applied opposite rudder, and just let go of the stick. By 1917, even inverted spins were tamed.

Vertical rolls

Vertical rolls had to wait until aircraft of sufficient power were available to sustain a suitable vertical line. Once this happened, the vertical roll was not a particularly adventurous discovery. RAF pilot Allen Wheeler flew the first recorder vertical roll in 1927.

Hesitation rolls

The hesitation roll was a pure display maneuver. By the mid 1920's, aerobatic competitions were commonplace and pilots were experimenting not to survive, but purely to impress. The fat fuselaged biplanes of the day had good knife edge performance, conducive to impressive hesitation rolls.

Outside loops

It took Jimmy Doolittle 435 hp to finally conquer the outside loop in 1927, a maneuver that had fascinated but eluded the aerobatic community since the days of Pegoud. The English were the first to succeed at flying a downward outside half loop (the English Bunt) in the early 1920's, but they didn't have the power to claw their way back up for a complete loop.

Rolling turns

The rolling turn was a contribution to the aerobatic community of Gerhard Fiesler of Germany. Fiesler was a pioneering competition pilot who first flew the maneuver in 1929. It is another purely competition maneuver and one of the hardest to fly. To develop it, Fiesler had to conquer sustained inverted flight, another "first" in his accomplished career.

For more aerobatic history, see "Flight Fantastic - The Illustrated History of Aerobatics," Annette Carson's landmark work on the subject.

11

Gyroscopic maneuvers

Gyroscopic maneuvers rely on the propeller's gyroscopic precession to induce a series of tumbles that takes its own course. In a gyroscopic maneuver, the pilot manipulates the controls to maximize gyroscopic precession and then hangs on for the ride as the aircraft assumes a mind of its own. Good advanced aerobatic pilots recognize when the energy in the tumble is spent, whereupon they recover in a coordinated fashion.

ORIGINS

During the 1950s, when the main international aerobatic event was the annual Lockheed Trophy held in Britain, the top pilots of what was then Czechoslovakia dazzled the world with a wildly tumbling maneuver that they dubbed the Lomcevak. The name is a Czech expression used to describe the unbalanced swaying of a person who has had too much to drink. It was the first gyroscopic maneuver to be introduced into the aerobatic arena. The Lockheed Trophy was judged to standards most reminiscent of today's four-minute freestyle program, with great emphasis on artistic impression. The Lomcevak was tailor-made for such an event.

The Lomcevak

The basic Lomcevak is usually flown on a vertical or 45° up line at a slow to moderate speed with full power. The pilot deflects the rudder to the right and applies full left aileron and forward stick (down elevator) simultaneously (the stick is simultaneously pushed in the left forward corner). What happens to the aircraft is graceful but difficult to describe. It tumbles end-over-end along all three axes on a projectile trajectory. As the gyroscopic force diminishes, the aircraft ends up in an inverted spin.

Difficulties in judging and restriction to the four-minute freestyle

Following the Lockheed Trophy's demise, the judging criteria shifted in modern competition aerobatics to evaluating the geometric precision with which standardized maneuvers were flown. This change in standards caused a big problem for the Lomcevak and other gyroscopic maneuvers in competition. Each aircraft type behaves differently when exposed to gyroscopic effect, making the standardization of a gyroscopic maneuver impossible. Also, the slightest control input differences can yield substantially different results even for the same aircraft. It is very difficult to fly gyroscopic maneuvers with the precision demanded of conventional aerobatic figures. These factors make it impossible to accurately judge gyroscopic maneuvers by conventional standards. They are therefore relegated to the four-minute freestyle, which is judged by different standards (addressed later in this chapter), more reminiscent of the Lockheed Trophy's rules.

The changes in judging standards notwithstanding, gyroscopic maneuvers remained very popular among top competition pilots and the race was on to outdo one another in the four-minute freestyle. Upon the Lomcevak's introduction, the aerobatic community started experimenting enthusiastically with the maneuver, and in the following years, several other identifiable gyroscopic figures were developed.

The torque roll

Charlie Hillard of the United States invented the torque roll some time before 1972, the year he became world champion. In the torque roll, which is also known as the rolling tailslide, the aircraft is pulled vertical and is continuously rolled to the left (with a right-turning propeller). The airspeed decays to the point where the vertical motion stops. The aircraft "hangs" on the propeller, which acts like the rotor blades of a helicopter, and the engine's torque effect keeps the aircraft rolling. Eventually the aircraft starts to slide back. At this instant, aileron is reversed to account for the change in relative wind, and the aircraft continues rolling to the left. To recover, the pilot closes the throttle and the aircraft swings down as it does in a tailslide.

The Zwibelturm

Eric Müller of Switzerland introduced the Zwibelturm, which means "spiraling tower" in 1974, the year he won the European championship. From a conventional vertical roll to the right on the up line, an autorotational tumble is induced that looks like a negative ascending spin, which is transitioned into an upright flat spin as the aircraft reaches the top of its trajectory and begins to descend.

The knife-edge spin

This maneuver is a contemporary of the torque roll and the Zwibelturm. It was first flown after the turnaround at the top of a hammerhead by simply continuing to hold left rudder and adding some forward stick (down elevator) and left aileron. Knife-edge spins are now performed on the up line as well as the down line.

Infinite variations

The introduction of the large six- and nine-cylinder monoplanes with their greater inertia and gyroscopic precession set off a boom in gyroscopic maneuvers, with the Russians in their big radial-engined machines leading the way. Along with this boom came a realization that the possible variations are infinite, and attempting to identify and name standard maneuvers is futile. The art of flying gyroscopic maneuvers evolved into the advanced aerobatic pilots' license to become test pilots. Today the pilots who dare to fly gyroscopic maneuvers all experiment with them until they each establish their own particular set of tumbles that are consistently repeatable and hopefully unique.

When you can do a tumble of your own design from this position, you've arrived!

FLYING THEM

We must begin this section with a giant disclaimer. You fly gyroscopic maneuvers at your own peril. To even think of trying this form of flying, you have to have an incredibly thorough understanding of every flying characteristic of your aircraft under every imaginable circumstance, and you must be super-proficient in every form of advanced spinning in general as well as in the aircraft you'll use for gyroscopic maneuvers. While you are learning, plan to start your maneuvers at a sufficient altitude to allow you to fully recover by the time you are at 3000 feet. And bear in mind that very few pilots in the United States today fly gyroscopic maneuvers, and most of them are highly experienced Unlimited competition pilots.

In gyroscopic maneuvers, the aircraft is affected by innumerable variables. Here are some of the most important ones:

- Aircraft weight
- Airframe shape
- Aircraft weight distribution
- Propeller speed
- Aircraft speed
- Angle of attack
- Angle of flight path to the ground
- Rate of control application
- Degree of control deflection
- Combinations of control inputs
- Power setting variations

The combinations are endless, the results unpredictable.

A good way to start learning to fly gyroscopic maneuvers is to first experiment with basic forms of tumbling. Set up the tumble and let it run its course. Try to carefully observe what happens as you hang on for the ride. When you can fully follow the tumble and can comfortably recognize a consistent pattern, it is time to start experimenting. Modify the control inputs at various stages and see what results. If you find a pattern you like, do it again and again. Eventually a recognizable and repeatable pattern will emerge.

Start with the basic Lomcevak on a 45° up line. When the aircraft slows to a moderate airspeed, apply full right rudder and simultaneously push the stick all the way to the left front corner. See what happens. When the aircraft settles into an inverted spin, recover and do it again and again.

Once you feel comfortable with this version of the Lomcevak, start varying it. Roll knife-edge before you start it; do it from a 60° up line and from a vertical up line. Experiment with changing the control inputs at various points in the maneuver.

The control inputs in gyroscopic maneuvers should be smooth, fluid applications. It is very important not to slam the controls to the stops. Slamming the controls is actually counterproductive because the relative wind can't keep pace with the change in the position of the control surfaces, which momentarily stall and are therefore less effective.

Another good way to begin learning gyroscopic maneuvers is to establish a 45° up line, roll inverted and aggressively but smoothly apply aileron, elevator, and rudder in the same "corner" (e.g., full left rudder and stick in the forward left corner). This simultaneous input of all the controls sets up a basic tumble. Let it take its course and recover. The characteristic of the basic tumble varies with airspeed and climb angle. When you are comfortable with the results of the basic variations, start experimenting with changing the controls.

Beyond this basic approach to beginning to learn gyroscopic maneuvers, you are really on your own. Trial and error will lead you to develop a style and bag of tricks that can set you apart from your peers. In a certain sense, gyroscopic maneuvers is the most exciting branch of advanced aerobatics because so much still waits to be discovered.

THE FOUR-MINUTE FREESTYLE

Gyroscopic maneuvers have become the staple of the four-minute freestyle. As its name implies, this event is time-limited and anything goes. You can fly any advanced maneuver, but if you don't include a healthy dose of gyroscopic maneuvers, you won't achieve a very high score. The four-minute freestyle is flown only in the Unlimited category on the last day of a competition. It is a completely separate event, judged separately from the regular competition. Its purpose is to retain the wilder, more freewheeling character of aerobatic competition that was lost when the emphasis in the sport shifted to competing in the context of established technical precision standards. It can be considered a sort of gala closing ceremony that provides the top Unlimited pilots a completely unconstrained, no-holds-barred opportunity to fly on the edge. Smoke is allowed, and music might soon also become an optional feature.

The four-minute freestyle has no elaborate judging criteria. The judges award scores in four categories and are not given any detailed technical guidelines. The four categories are

- Originality
- Versatility
- Execution
- Harmony and rhythm

The judges watch and score the entire event. Following the event's completion, each judge can review all the scores he or she gave and readjust them as he or she sees fit. The intent is to provide another opportunity to make the scoring consistent in view of the field's overall performance and the absence of detailed judging standards.

At large events, the freestyle is not open to just anyone. Usually the top third of the field is invited to fly, and participation is not mandatory.

Training for it

The four-minute freestyle is a grueling event. As one world-class competition pilot said, "It is difficult to get through four minutes of flying out of control without running out of altitude, airspeed, and ideas." To do well in the four-minute freestyle, you have to put a tremendous effort into devising and practicing a sequence of five or six complex gyroscopic maneuvers. Most pilots simply don't have the time to prepare equally for the regular competition and this event and choose to put most of their time and energy into training for the regular competition. One alternative that has been informally discussed that might make the event more accessible to more pilots is to cut it to three minutes. Ultimately how much effort you put into training for the four-minute freestyle is a matter of personal choice. But you can be sure that if you do give it an honest try, you'll experience some of the wildest, most punishing, and most rewarding form of aerobatics on the planet.

Trends

There is a current trend to establish more competition opportunities for this style of flying. The fact is that spectators find the exuberant showmanship of the four-minute freestyle more exciting to watch than the regular competition (especially the repetitive displays of the compulsory known and unknown sequences). The hope is that events founded on the four-minute freestyle will turn aerobatics into a true spectator sport. Breitling was the first to experiment with the concept when it sponsored the FAI-sanctioned Breitling World Cup of Aerobatics during the early 1990s, which drew its competitors from the top finishers at the world championships. Similar to world-cup skiing and Formula car racing, several events were held during the year throughout the world, and the overall winners for the year were determined by a point system. Efforts are underway to establish an FAI-sanctioned grand prix of aerobatics along similar lines, supported by multiple sponsors.

World champion Xavier de L'Apparent's four-minute freestyle tips

Scoring the four-minute freestyle is very subjective. The tendency is to favor the most dynamic and, above all, the most spectacular flights. In spite of that, the judges do not all appreciate the same style of flying. Some like very flowing flights, others like a more staccato style.

To be appealing, a program must have rhythm. Keep it moving! The judges musn't be allowed to fall asleep. The movements that slow down the tempo are most often slow rotations repeated again and again, long lines that go on forever, or long lines before and after figures.

The rhythmic aspect of a program can easily be improved in several ways: by dressing up the straight lines with rolls or snap rolls; by alternating rapid and slow movements; by alternating the direction of rotation; and by limiting the number of figures used to regain altitude.

Originality is very important and, notwithstanding the other prescribed criteria, will always be the deciding factor between the flights of two different competitors. The ideal thing would be to invent an entirely new figure not based on anything already known. But such a discovery is not achievable without a great deal of thought, training, and, above all, inspiration.

If one has the time and the means, it is advisable to work on the four-minute freestyle just as one does with a known or free program. First make a selection of all the figures that you can reliably execute, and then link them together intelligently. When you have a well-structured sequence, you can learn it by heart.

With a winning sequence, there is no place for a figure, no matter how spectacular, that ends by being "botched up" (indistinctly). Non-catalog figures also have to be as clean as possible. When the aircraft does not react as planned, it is essential to recover the situation by coming out of the figure on a precise heading on a 45° line, vertical, or in level flight.

In spite of everything, subjectivity plays a large part. After all the comments I have been able to gather in the course of my aerobatic career, in the opinion of the judges, seasoned spectators, and the uninitiated, one point stands out clearly every time: the spectacular. People are impressed not so much by the figures themselves but by what the aircraft can do. To be spectacular vis á vis the judge, the pilot must be able to describe totally unexpected trajectories with the aircraft, but with precision.

Xavier de L'Apparent is the 1994 World Aerobatic Champion, 1994 French National Champion, and winner of the 1994 Breitling Cup. He is the owner and operator of Magic Voltige, an international school of advanced aerobatics, and operator of the Breitling Academy. These comments are presented with kind permission from his book, The Aerobatic Four Minute Freestyle, which can be obtained from Magic Voltige Publications, 19 domaine Bois de Jarcy, 91480 VARENNES-JARCY, France.

Xavier de L'Apparent about to take off on the flight that won him the 1994 World Championship.

12

Advanced
sequence composition

One of the more challenging aspects of being an advanced aerobatic competition pilot is having to design and perform your own freestyle sequences. Freestyles are a requirement in advanced-level competition in every country where competitive aerobatics is flown, as well as in international events. In the United States, freestyles are an option in Sportsman class and are a compulsory requirement starting with the Intermediate class. In this chapter, we discuss all the considerations that go into designing sequences. We take you through examples of sequence composition techniques. And we present the 1996 IAC known sequences from Sportsman to Unlimited to demonstrate sequence design principles for different levels of difficulty.

Three factors have significant influence on sequence composition:

- Aircraft performance
- Pilot skill level
- Flying style

The objective of sequence composition is to devise a sequence that maximizes the pilot's potential score. This goal can only be accomplished by carefully taking into consideration the influence of these factors.

The FAI catalog provides so many maneuver permutations and difficulty levels that there are practically unrestricted opportunities to custom-design a freestyle sequence that builds on the strengths of your aircraft and your piloting skills and minimizes any weaknesses.

AIRCRAFT PERFORMANCE

We talk of an aircraft's performance relative to its competitors. Yesterday's high-performance aircraft are today's middling performers as new, improved models

Rick Massegee doing what he did best.

IAC/Dave Gustafson

enter the market. Yet with cleverly custom-designed sequences, a good pilot can remain very competitive in freestyle events for a very long time in an aircraft that is slowly being overshadowed in performance by newer models. There are also minor differences in performance among the most advanced aircraft of the day, which the perceptive pilot seeks to exploit by resourceful sequence design.

The pilot of an aircraft that is no longer on the cutting edge of performance can be at more of a disadvantage in the known and unknown sequences, having no control over designing them. This factor makes a well-designed freestyle all the more important. It provides an opportunity to regain any ground lost to disadvantageous sequences in the other events of a competition.

Two aspects of aircraft performance have significant bearing on sequence composition: vertical performance and roll rate. Aircraft with lower vertical performance are also generally the ones that have the lower roll rates.

Low vertical performance and roll rate

Let's take vertical performance first. Vertical performance is constrained by engine power and drag. Any maneuvers on the vertical line add to drag and further cut into vertical performance. Therefore, in aircraft with low vertical performance, vertical lines should be kept to a minimum, and maneuvers along them should be restricted to lower-drag maneuvers, such as the half roll. Avoid vertical snaps and full vertical rolls.

The best strategy to overcome unexciting vertical performance is to design a sequence in which the high-drag, complex components of the maneuvers are in the horizontal plane. The typical freestyle sequence for aircraft that fall in this group contains a lot of looping maneuvers loaded down with K (difficulty factor) in the horizontal plane. Look at Fig. 12-1, which we first encountered in Chapter 1. The basic maneuver is a half loop worth 6K, which allows low-drag vertical maneuvering. The sequence is loaded with additional K at both horizontal segments to maximize the potential score. At the bottom is a four-point roll (11K). At the top is a 3 of 2 (12K) followed by a double snap (17K). The total maximum score for this maneuver is 46K, compared to 6K for the simple half loop.

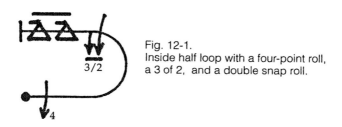

Fig. 12-1.
Inside half loop with a four-point roll,
a 3 of 2, and a double snap roll.

In Unlimited competition, the maximum total score for the entire sequence is set at 446K. If the sequence has 14 maneuvers, the average K per maneuver is 32K. The maneuver we designed is well in excess of this average.

Now consider a low roll rate. This characteristic is more pronounced in comparison to higher-performance aircraft at low speeds. At high speeds the difference is much less. Therefore, it is better to insert most rolling maneuvers at the high-speed segments of the sequence, such as the bottom of a loop.

Why, then, you may ask, is there a roll in our example? It is there because you need the high K to achieve a sufficiently high potential score for your sequence. You can't always avoid inserting a roll where you don't want one. If you have to place rolls at both the high-speed and low-speed segments of a maneuver, you should match the roll rates to make the maneuver appear well-balanced. A good way to do this is to offset a continuous roll at the slow segment with a point roll at the fast segment, as we have done in this example.

A lack of roll rate can also be disguised by making frequent use of hesitation rolls. A roll through 90° or 45° followed by a pause is very effective in cutting down the visible difference in roll rate compared to a higher-performance aircraft.

On the slow horizontal segments of a sequence, such as at the top of a loop, the best maneuvers for an aircraft with a low roll rate are variations of snaps. The required entry speed is moderate, and autorotation takes care of the roll.

Look at Fig. 12-1 again. You can see the four-point roll at the bottom and the double snap at the top.

High vertical performance and roll rate

High vertical performance and roll rate do not provide an automatic advantage over lower-performance aircraft. If misused, high performance and roll rate can be a hindrance. Let's start with vertical performance.

The greatest advantage to high vertical performance is that it allows the pilot to fly the sequence at a lower altitude because of the higher energy available. The result is a tighter, better presentation.

If, however, an aircraft has high performance, it must use it. The higher-performance competition aircraft, such as the Extra 300S and the Sukhoi 31, are bigger and heavier than, for example, the Extra 230 and the Pitts, and they look slower at similar speeds. To present well, they must be flown faster (at least 200 mph or above for the cited aircraft). Many pilots transitioning into higher-performance aircraft don't realize this and struggle with lackluster presentation until they figure it out or a good coach tells them.

Look at Fig. 12-2. It is a pull pull hammerhead with a full roll up and a double roll down. The maneuver makes good use of vertical performance. The basic hammerhead has a score of 17K. The full roll up is 12K, and the double roll down is 12K, for a total score of 41K.

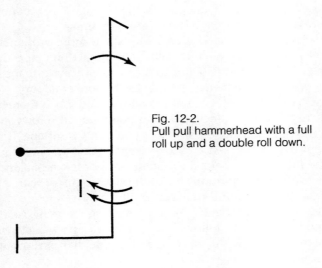

Fig. 12-2.
Pull pull hammerhead with a full roll up and a double roll down.

Now let's consider roll rate. In today's environment, a roll rate is considered high if it is 360° per second or more. The top aircraft have roll rates in excess of 400° per second. Rolls have a spectacular effect on a sequence. The big advantage of a high roll rate is that the sequence can be creatively packed with variations of rolls. Full rolls, multiple rolls, hesitation rolls, and hesitation rolls with rotation reversal are just a few of the variations.

The only rolling maneuvers you should consider avoiding in an aircraft with a high roll rate are multiple rolls on the vertical up line. The roll rate is so high that the horizon is a blur, and it is very difficult to precisely stop any roll beyond a single roll. Certainly it is not worth the risk of losing points. Stick to half rolls and single rolls on the vertical up line.

Look again at Fig. 12-2. It is about as difficult to fly and is worth as many points as the maneuver in Fig. 12-1 designed for aircraft with a lower vertical performance and roll rate. Fly either one well with the aircraft best suited for it, and you will score equally high. Do bear in mind that you have more flexibility in designing sequences for aircraft with good vertical performance and roll rate because they tend to perform and present equally well in the horizontal plane.

PILOT SKILL

You need to make an honest, detailed assessment of your flying skills when you design a freestyle sequence. It is obviously pointless to design a sequence that is beyond your capabilities, but it might also be detrimental to compose one that takes you to the edge of your abilities. Flying a competition sequence that takes you to the edge of your abilities allows a great potential for making mistakes and winding up with a dismal score.

Aerobatics requires a building-block approach. The foundation of your skills must be rock-solid, and you must be able to perform to perfection at a particular level before you move on to the next. If you push yourself too hard, you run the risk of not acquiring the solid grounding to do truly well in top-level national competition. Attempting to do it retroactively more often than not proves to be a long and frustrating process.

Design your sequences with maneuvers that you can do well and with great confidence. Concentrate on maximizing the sequence's potential score within your flying abilities. Then go and fly your heart out. Push yourself to the limit by attempting to do better at what you do well. When you reliably achieve high scores, it is time to raise the ante. In the same vein, stay in a competition class until you consistently finish in the top rankings (at the regional or national level, depending on your goals) before you move up to the next level.

Some people hold the opinion that flying sequences that are at the edge of your capabilities or even slightly beyond them is the way to gain experience and move forward. We think this approach is counterproductive. Besides missing the nuances of what you are doing wrong because you are in over your head, you will become frustrated by making a lot of errors and scoring low and will run the risk of becoming demoralized. You will be struggling to just get through your maneuvers, and you will have a difficult time compensating for the wind and concentrating on presentation.

There certainly is a place for frequently flying on the edge. It is the practice session. That's where you should push yourself to the limit to reach for the next level. And when you can comfortably fly a tougher maneuver to average standards, it is time to put it in the sequence.

One sure way to design a good freestyle sequence is to include only maneuvers that you like to fly. If you enjoy them, you will probably do them well. After all, the most important reason for flying aerobatics is enjoying it.

PRESENTATION STYLE

Your flying style and the presentation you wish to achieve have to be factored into designing your sequences. Do you prefer a long, lyrical sweep peppered with dainty pirouettes through the sky, or are you a fan of aggressive power aerobatics? Would you like to perform ballet, or would you like to rock and roll?

To a large extent, this choice is a personal preference, but to some degree it is also a function of the flying characteristics of your aircraft. Some aircraft are more suited to perform a ballet than to rock and roll.

Ballet

You can take several steps to give your sequence a graceful lyrical style. Draw long, sweeping lines and many looping maneuvers. Add a large number of continuous rolls. Do many hammerheads, tailslides (Unlimited only), and humpty bumps.

Rock and roll

Rock and roll is brute power. This kind of performance calls for a high-speed, highly energized sequence. Multiple snap rolls galore are the staple maneuver. Peter Bessenyei, the world-class Hungarian competitor, aptly dubbed this style the contest of a thousand snap rolls. Super-crisp point rolls with roll direction reversals also contribute to the desired effect.

SEQUENCE REVIEW

To get an idea of how sequences are designed and how the level of difficulty varies from category to category, take a look at Figs. 12-3 through 12-6. They are the Sportsman through Unlimited knowns for 1996. Compare the sequences, think about the reasons the maneuvers are sequenced as they are, and identify the wind correctors. We included the Sportsman sequence, consisting entirely of basic maneuvers, to illustrate the difference in level between basic and advanced aerobatics.

Sportsman

1. Half Cuban with 2 of 4 to upright
2. Immelman
3. One turn inside spin
4. Hammerhead
5. Reverse half Cuban
6. Loop
7. Half Cuban
8. Slow roll
9. 270° turn
10. 90° turn
11. Two-point roll

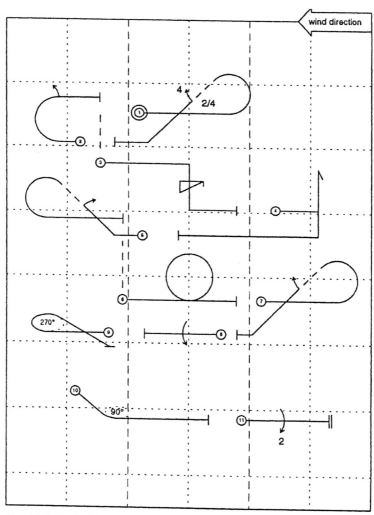

Fig. 12-3. Typical Sportsman sequence.

Intermediate

1. Loop
2. Hammerhead with a ¼ roll up and a ¼ roll down
3. Half inside loop to inverted
4. Half roll to upright
5. 1¼ inside spin
6. Pull, pull, pull humpty bump with a ¼ roll down
7. Reverse half Cuban
8. Inside square loop
9. Half Cuban
10. Immelman
11. 90° turn
12. Snap roll
13. 270° turn
14. Split S
15. Slow roll

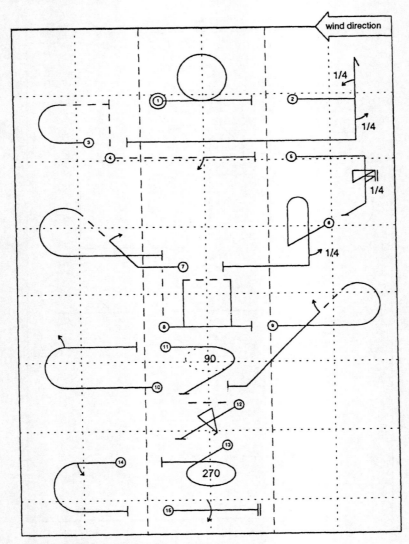

Fig. 12-4. Typical Intermediate sequence.

Advanced

1. Hammerhead with ½ roll up
2. Pull, push, pull humpty bump with half roll up
3. Reverse half Cuban
4. Half inside snap roll on 45° up line, layout to inverted
5. 1¼ turn inverted spin, pull to upright
6. Pull, pull, pull humpty bump, with ¼ roll up
7. Inside outside horizontal eight
8. Immelman
9. Two point roll on 45° downline
10. Hammerhead with ½ roll down pushing out to inverted
11. Slow roll from inverted to inverted
12. Half outside loop to upright
13. Rolling circle with three rolls to the inside
14. Split S
15. ¼ vertical roll, layout to inverted
16. ½ square loop own
17. ¾ vertical roll with cap off upright

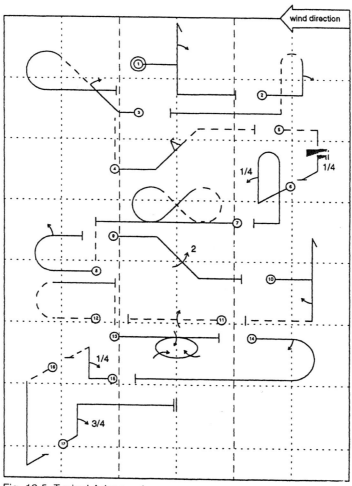

Fig. 12-5. Typical Advanced sequence.

Unlimited

1. Pull, pull, pull humpty bump with 1¼ roll up and 1¼ inside snap roll down
2. 1¼, ¾ rolls in opposite direction on 45° up line, ⅝ outside loop to vertical down with full vertical roll, pull out upright
3. Shark's tooth, full vertical roll up, ½ roll to upright, followed by double inside snap in the opposite direction, pull out upright
4. 1½ vertical roll, capping off upright
5. Push down to 45° line, two point roll, followed by ⅝ loop to inverted
6. Rolling circle with four rolls to the inside, starting inverted and finishing inverted
7. Push pull hammerhead with a 1¼ outside snap roll
8. Pull, push, pull humpty bump, 2 of 4 on the up line with ¾ roll down
9. Half vertical roll followed by ¾ inside loop
10. Shark's tooth with 2 of 4 on 45° up line
11. Hammerhead, 3 of 4 going up and ¼ roll down
12. 45° humpty bump, 4 of 8 on the 45° up line 1½ inside snap down, push to inverted
13. ½ roll, followed by ½ inside loop with a full roll from inverted at the top
14. 2 of 4 from inverted to upright, followed by a double inside snap roll in the opposite direction

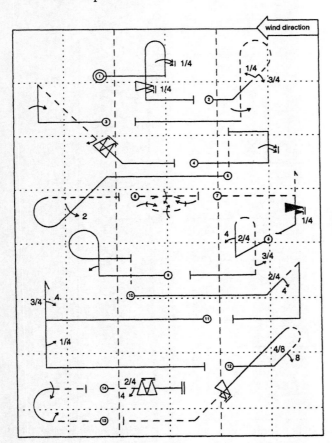

Fig. 12-6.
Typical Unlimited sequence.

Lastly, Fig. 12-7 is Mike Goulian's freestyle sequence for 1995, the year he first became Unlimited U.S. National Champion. If you already have your FAI catalog, see if you can work your way through it. Fig. 12-8 is a judge's score sheet for the sequence.

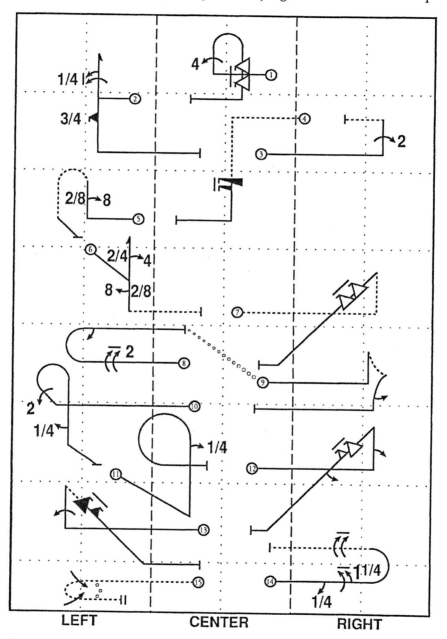

LEFT   CENTER   RIGHT

Fig. 12-7. Freestyle sequence flown by Mike Goulian when he won the 1995 Unlimited U.S. Nationals.

No.	symbol	catalogue no.	K	total K	grade	remarks	No.	symbol	catalogue no.	K	total K	grade	remarks
1		8.1.1 9.4.1.4 9.9.5.8	13 15 17	45			13		1.16.2 9.1.1.4 9.10.4.6	13 12 16	41		
2		5.1.1 9.1.1.5 9.10.5.3	17 14 13	44			14		7.1.1 9.1.3.1 9.1.3.5 9.1.3.8	6 2 9 12	29		
3		1.7.1 9.2.1.4	9 13	22			15		2.6.2	14	14		
4		4.8.6	15	15			16						
5		8.3.1.1 9.8.1.1	15 7	22			17						
6		5.1.3 9.4.1.2 9.8.5.1	18 9 3	30			18						
7		1.17.1 9.9.4.8	16 17	33				POSITIONING			26		
8		7.2.1 9.2.3.8 9.1.3.2	6 15 4	25				FIGURE TOTAL K =			420		
9		6.2.1 9.1.5.2	15 4	19				INCLUDING POSITIONING =			446		
10		8.57.1 9.2.2.4 9.1.5.1	12 11 2	25				JUDGE'S NAME					
11		8.33.1 9.1.1.1	11 6	17				JUDGE'S NUMBER					
12		1.16.3 9.1.1.2 9.9.4.6 9.1.4.2	14 8 14 4	40									

INTERNATIONAL AEROBATICS CLUB SCORESHEET — CATEGORY **UNLIMITED** — pilot's number

IAC

Fig. 12-8. Judge's score sheet of freestyle sequence flown by Mike Goulian when he won the 1995 Unlimited U.S. Nationals.

Diana Hakala
on freestyle construction

My suggestions have an underlying assumption: The "best" freestyle is the simplest to fly and the easiest for the judges to award the highest score. The simpler the sequence, the less room for error by either you or the judges. A simple sequence can achieve maximum K just as well as a complicated sequence. Always use the maximum K available. Many a contest has been won or lost by one point.

BASIC DESIGN

1. Put a maximum of three maneuvers upwind and two downwind on the X axis before inserting a wind corrector (crossbox maneuver) on the Y axis.

2. Put the wind correctors on the upwind side. You'll be pushed into the box while you fly them.

3. Plan to use the maximum number of maneuvers allowed. This technique averages out your total points over more maneuvers, reducing the damage from an error in any one maneuver.

4. Fly the first maneuver center-box, and make it an exciting figure. It is easier to center the first maneuver and best to get an extra-difficult one out of the way up front.

CHOOSING MANEUVERS

1. Minimize 45° lines. The 45° line is difficult to judge accurately, especially when flown at high altitude or close to the judges.

2. Use ½ and ⅝ loops rather than complete or almost-complete ones. It is easier to see problems with the latter because the airplane draws a more complete geometric image. A complete loop has to be done center-box, where errors are especially easy to see.

3. Minimize the number of rolls in rolling turns. They are difficult to fly as well as to judge accurately.

4. Minimize negative maneuvers. On a 100° F day, you'll be glad you did.

5. Minimize the K on tailslides. It is difficult to get a good score on tailslides because you have to "cheat" to make sure you swing in the right direction.

PLACING MANEUVERS IN THE SEQUENCE

1. Spins, hammerheads, and tailslides work best on the upwind side. The risk of a line error is greatly reduced.

2. Finish downwind maneuvers into the box. For example, have the top of a humpty bump head toward the center, into the wind. This technique takes the guesswork out of how far out to start the maneuver in varying wind conditions. For the same reason, you are better off with a half Cuban over a reverse half Cuban downwind.

3. Enter spins into the wind. A spin entered downwind looks forced.

4. Keep vertical rolls simple on the way up, 180° and 360°. Anything more complicated, such as the 1½ roll, is much easier to stop accurately on the downline.

5. Avoid placing hammerheads and humpty bumps in the center-box, especially with a 90° or multiple turn, which makes it very easy to see if your wings are level. If you ¼ roll on the way up, the timing of the pivot is also easy to critique.

6. Humpty bumps and ⅝ loops are good crossbox maneuvers.

When you've completed your sequence and added up the K, you'll probably find that it is off by a point or two. You can fine-tune the K by adjusting rolls and turns. For example, you can change a ¼ roll to a hesitation roll or a 90° turn to a 270° turn to adjust K.

Diana Hakala is a highly experienced Unlimited competitor and a U.S. National Team member. She flies a Staudacher S300.

Diana Hakala flying her pretty Staudacher S300.

13

Presentation

When you first begin to fly competition aerobatics in the aerobatic box, you should be pleased with yourself if you are able to get through your sequences within the confines of the box. However, to the observer on the ground, aerobatic maneuvers can look quite different when flown in different segments of the box. And if that observer is a judge, appearance is especially important. As you move up in the ranks, you have to become increasingly more proficient at restricting the aircraft's flight path to a certain inner segment of the box, which we call the performance zone, to achieve the best presentation of your sequence.

THE IMPORTANCE OF GOOD PRESENTATION

By presentation, we mean the appearance of a sequence to the judges. A maneuver can be flown perfectly, but if it is performed in an area of the box in which it is difficult for the judges to accurately evaluate it, the contestant's score is likely to suffer. Good presentation is important not because you are specifically judged on it, but because it makes the judges' job of accurately evaluating each figure easier, thereby increasing your chances of achieving a high score.

The accuracy of a maneuver's geometric shape is one of the most important criteria by which it is judged. The problem in accurately judging a maneuver arises because the judges are viewing a one-dimensional geometric shape in a three-dimensional world. The angle from which a one-dimensional shape is viewed distorts its appearance to the viewer.

One example of this phenomenon is viewing a loop from different vantage points. A loop that is flown very close to the judges is almost impossible to judge because of the viewing angle (Fig. 13-1), even though it is perfectly round. A tight and perfectly round loop flown low and in the center of the box is viewed by the judges at a much shallower angle and is therefore much easier to judge. Its appearance to the judges more closely resembles the aircraft's actual flight path.

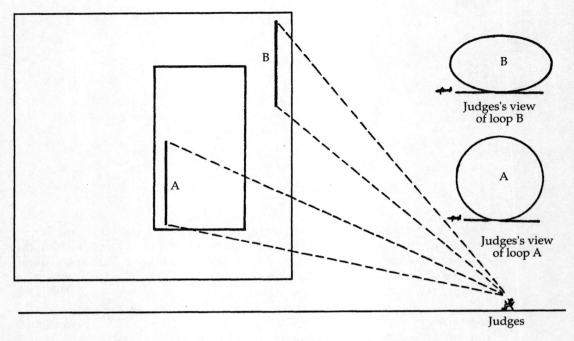

Fig. 13-1. The same loop in different parts of the box looks different to the judges.

You can readily see this effect by conducting a simple experiment. Draw a perfect loop on a sheet of paper and hold it at arm's-length directly in front of you so that the paper is perpendicular to the floor and your arm is horizontal to the floor. Your arm represents your line of sight. The loop appears perfectly round. Now raise the paper by rotating your arm up at about a 75° angle and keeping the paper perpendicular to the floor. At this viewing angle, the loop appears elliptical.

Lines can also present a problem. The most difficult line to accurately evaluate is the 45° line. Flown close and high, it looks much steeper than 45°. Off to the side and far away, it can appear much shallower. Vertical lines flown at the edge of the box are harder to evaluate than center-box because viewing the aircraft obliquely, rather than in profile, makes it difficult to tell if it is perfectly vertical, positive, or negative.

Judges are trained to compensate for distortion, but the more extreme the distortion, the more difficult it is to accurately account for it. There will always be some distortion, but it can be reduced by skillful presentation. The objective of presentation is to restrict the performance of a sequence to a viewing angle that maximizes the accuracy of each figure's appearance to the viewer. This goal is accomplished by flying your sequence inside the performance zone.

THE PERFORMANCE ZONE

The performance zone is the area of the box in which your maneuvers appear most true to the judges. It is not an official airspace but one that is informally determined by trial and error to achieve the desired goal, taking into consideration the confines of aircraft performance and flying skill. Vertically, the performance zone is between 500 feet and 2500 feet AGL (the top of the box is 3300 feet AGL). The floor is obviously higher for those classes that have a higher-than-500-foot floor for the regular box. The performance box's width is 500 feet less on each side than the official box, and its depth is from centerline to three quarters of the way to the edge of the official box, toward the judge's line (Fig. 13-2).

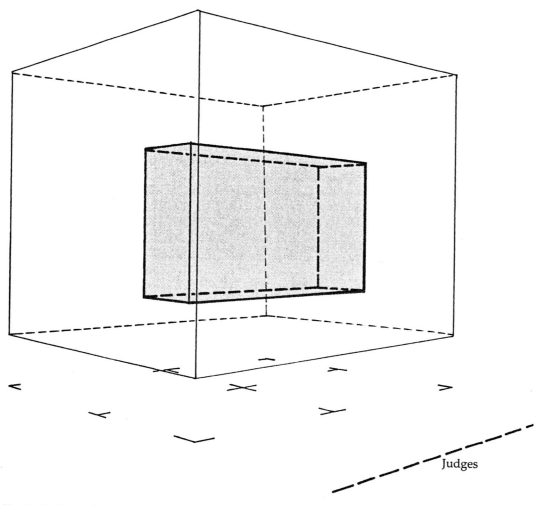

Fig. 13-2. The performance zone.

In addition to improving presentation, learning to fly sequences in the performance box has another benefit. You will have a generous buffer zone between the formal boundaries of the box and the area in which you generally fly, reducing the chances of costly boundary violations (Fig. 13-3).

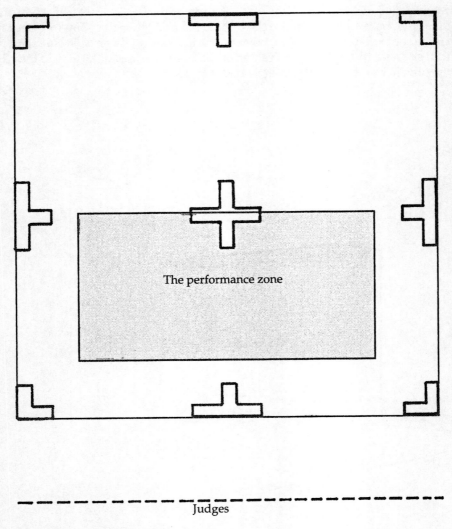

Judges

Fig. 13-3. The performance zone, pilot's view.

John Morrissey, U.S. team coach, on competition training

It is interesting to compare how different countries prepare their national teams for international competition. From my six weeks of experience at the Soviet/Russian training camps in Borki, Russia (latitude 5°30' north, closer to the North Pole than Juneau, Alaska), I learned that it is not necessarily how many hours you fly, but how you spend your time that counts.

The average Russian team pilot flies approximately 40 hours per year during four training camps that last about three weeks each. During these camps he or she flies three to four 17-minute sorties per day in an unmarked box. Approximately 60 percent of these sorties are critiqued by one of their trainers. The remaining sorties are flown in outlying "zones" to work out maneuver techniques. Videotaping is not used at these camps. The pilots live in a hostel on the field. The entire purpose in life is to win medals in international competition. They have no other distractions. During their considerable off time, they are required to stay in good physical shape and are also encouraged to maintain a healthy, well-balanced mental attitude. As Russia and other former Soviet republics go through their current period of profound change, it will be interesting to see how their aerobatic opportunities, training, and attitude will be affected.

In my own training camps, which I conduct for all categories of competition, I feel the best results are obtained by using a marked box and providing videotaping with the voice critique/coaching overlaid on the videotape. I purposely do not align the box with any natural landmarks to force the pilots to learn to fly the box instead of outlying references, which might not be available in critical international competitions. A good example is Le Havre, site of the 1992 World Aerobatic Championship, where two-thirds of the horizon was the English Channel. At other international competitions, the alignment of the box has been changed on a daily basis.

As far as training techniques are concerned, I do not believe in continual practice of sequences. My experience has shown this method to be an uneconomical and fatiguing way to practice. When my son won the first contest he entered, he had flown the sequence in its entirety only once. He had, however, flown the first three maneuvers in the contest's marked box over 50 times. The rest of the sequence was broken down into three to four maneuver segments for *box* practice.

I believe in exercises that continually refer to the box and "work" the wind for their successful completion. Examples of this technique include the "four corners," where wing slides (hammerheads) are flown as close to each corner as possible without letting the existing wind displace the pilot from the competition zone.

I have never seen much progress made in practice flight after the 15-minute mark in the flight, especially if it involves an Advanced or Unlimited sequence. Overpracticing, in my view, is almost as big a problem as underpracticing. I have never seen a brilliant sequence on the third flight of the day. Try to remember that the coach almost always sees the fatigue set in before the pilot feels the effect of overpracticing.

Finally, the pilot should remember that in aerobatics, there are no messiahs. Good training is wonderful, and several points of view can be very helpful; however, in the final analysis, the coach cannot create the final performance. Only the individual pilots are capable of this, and to do so they must use the training received to allow their own style, rather than the trainer's, to dominate the performance they present to the judges.

John Morrissey is the coach of the U.S. National Team, a post he has held for many years. He also operates his own aerobatic training school and gives individual coaching at all levels.

John Morrissey, U.S. team coach, and his wife and U.S. team member, Linda Meyers Morrissey, at the Worlds.

14

Handling the wind

The wind can be a nuisance in aerobatic competition flying, callously foiling your best-laid plans. It mutates the geometrical shape of your maneuvers, wreaks havoc with your ground alignment, pushes you where you don't want to be, and might even unceremoniously eject you from the box. Contest officials may stop the competition when the wind speed reaches 20 knots, but they can waive this rule, and for advanced-competition aerobatics, they generally do.

Many competition pilots place little emphasis on maximizing their control over the wind. They seem to be content with learning to merely stay within the box's confines on windy days. But one pilot's nemesis is another pilot's opportunity. Put a lot of effort into learning to tame the wind to minimize its effect on your all-important presentation, and you will be handsomely rewarded in the competition standings.

Handling the wind needs to be discussed from two perspectives:

- Correcting for the wind's effect on the geometrical shape of the individual maneuvers
- Correcting for the wind's effect on your position in the box throughout the sequence

WIND-CORRECTING FOR
MANEUVER ACCURACY

The wind distorts the geometrical appearance of maneuvers if they are flown without any correction for its effect. A lot can be done to maintain the geometrical accuracy of looping maneuvers. However, no wind correction can be applied to vertical maneuvers to maintain maneuver accuracy (as you'll see, some preemptive steps can be taken for positioning). On the vertical line, the aircraft's attitude has to be perfectly vertical (Fig. 14-1). Any correction would require it to be positive or negative, which would result in a loss of points. We can adjust the spot in the box where the aircraft is pulled vertical, but once it is vertical, so it must remain, pushed (as it will be) by the wind. The judges won't penalize you if the line you draw isn't perfectly vertical due to wind. They will penalize you if the aircraft's attitude is not perfectly vertical.

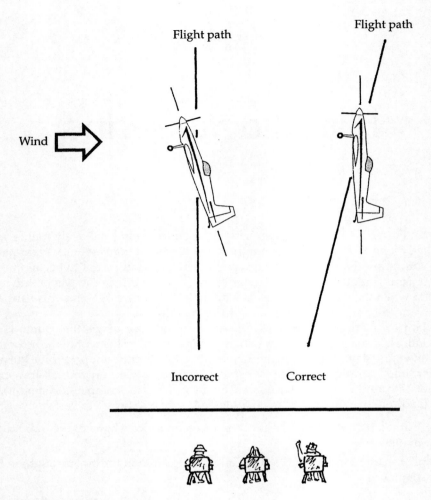

Fig. 14-1. Proper vertical alignment in wind.

Looping maneuvers are another story. Here all kinds of corrections can be (and need to be) made to keep them perfectly round. But before we explore these techniques, you need to perform a series of exercises to learn to gauge the wind's effect on your aircraft.

Go out on a windy day and find a straight stretch of highway or similar landmark that is perpendicular to the wind direction. Fly into the wind. Pull vertical over the highway and do a hammerhead. As you pivot at the top, look at the ground and note where you are relative to the highway. Repeat the exercise five or six times. Now go and perform the same exercise again, but this time do it downwind five or six times. Go up on days when the wind strength is different, and repeat the exercise. When you can reasonably predict where you will be at the pivot relative to the highway, you have begun to develop a feel for the effect of wind on your aircraft. You are ready to tackle wind-correcting for looping maneuvers.

Loops can be flown into the wind or downwind. Let's first see what happens to an uncorrected loop flown upwind and downwind in a 20-knot wind (Fig. 14-2). Upwind, the first half of the loop gets compressed by the 20-knot headwind. As the aircraft floats over the top, it now has a 20-knot tailwind and gets pushed downwind. It is further pushed downwind on the second half and completes the figure a considerable distance from where it began, resulting in an elliptically shaped maneuver.

If the loop is flown downwind, the aircraft gets pushed back on the pull and faces a 20-knot headwind just as it runs out of energy at the top. On the second half of the loop, the headwind compresses the figure. The aircraft is again pulled out downwind of its starting point, and the loop's shape is again elliptical.

Let's look first at correcting the loop upwind. Take a look at Fig. 14-3. During the first stage of the pull (Zone 1), you need to ease off the normal acceleration. You "drive" the airplane into the wind to establish the figure's circular shape that the wind would otherwise begin to compress. During the second stage (Zone 2), you pull more G than normal, speeding up the flight path toward the top of the loop. If you pulled normal G here, you would be carried too far downwind by the time you reached the top of the loop. The float over the top should be of slightly less duration than normal (Zone 3). On the downwind half immediately following the float (Zone 4), you need to pull about ½ to 1 G more than normal to continue maintaining the figure's circular shape. On the pull-out (Zone 5), you again need to ease off slightly, to "drive" back to the starting point of the loop. This last segment is where most pilots are inclined to continue to pull hard, perhaps influenced by the rapidly building speed. This tendency ends up "hooking" the end of the loop instead of giving it a nicely rounded finish (Fig. 14-4).

When you fly a loop downwind, you pull harder than normal initially, float for much longer than normal over the top into the headwind, and ease off the Gs in comparison to normal loading following the float.

Because it takes an agonizingly long time to drive into the wind in the float over the top, and it is easier to shape looping figures when you have more speed, loops, Immelmans, half loops, half Cubans, and other variations of the loop should be flown into the wind whenever possible.

WIND-CORRECTING FOR POSITIONING

The basic objective of positioning is to stay within the performance zone for best presentation. Adjusting for the wind to accomplish this goal requires you, depending on circumstances, to maintain position over the ground or move upwind.

A direct headwind or tailwind presents a relatively simple problem. You have to spend more time trucking upwind and less time downwind as you transition from one maneuver to the next. It is all timing. Beyond that, follow the principles of maintaining the geometrical accuracy of individual maneuvers, and you'll do fine. However, there is rarely a direct headwind or tailwind. It is much more likely that a crosswind component will rear its ugly head, and then the situation becomes more challenging.

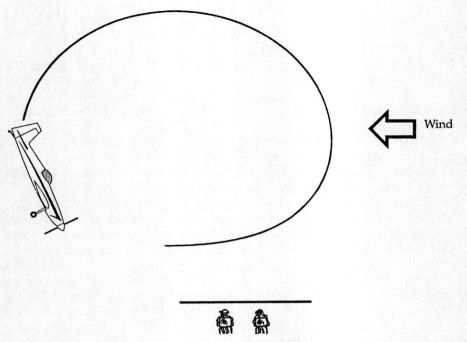

Uncorrected loop into the wind

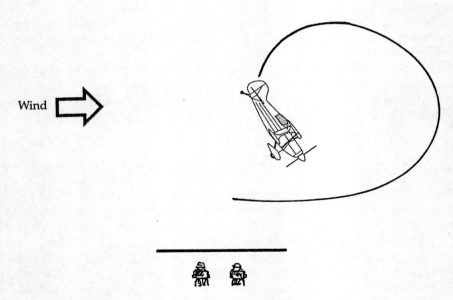

Uncorrected loop downwind

Fig. 14-2. Uncorrected loops in wind.

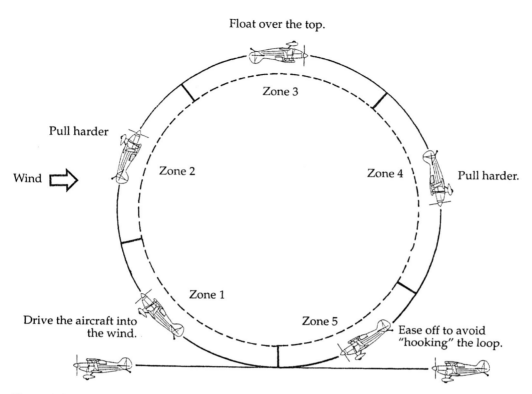

Fig. 14-3. Wind-correcting the loop.

A crosswind wind can push you off centerline (the X axis) in the box to varying degrees, depending on the crosswind component (wind direction, wind speed, and aircraft speed). Confusing as this problem can become, there are several ways to deal with it. The simplest and most common method is to insert maneuvers known as wind correctors in your sequence, designed to reposition you at strategic intervals during your sequence. Two other techniques are more challenging but enable you to more effectively minimize the wind's effect on your presentation by keeping the aircraft on or close to centerline throughout the sequence. One involves constantly pointing the aircraft's nose into the wind. The other utilizes sideslip, which is interesting given the admonishment against having any sideslip in vertical rolls and tailslides. Let's take a closer look at all of them.

Wind correctors

A wind corrector is a maneuver along the Y axis of the aerobatic box. It should be inserted into a sequence after approximately every five maneuvers. The aircraft starts the sequence and is flown through each maneuver without adjustment for

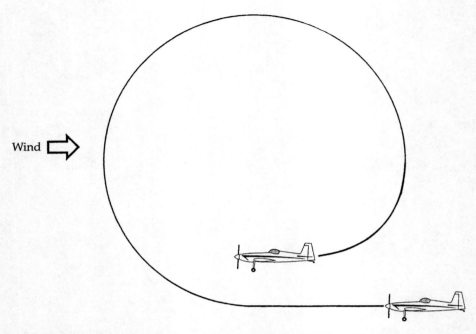

Fig. 14-4. Hooking the loop in wind due to a failure to ease off the stick at the end.

the wind. It begins to be displaced by the wind. The idea is that it can get through five maneuvers prior to being so far out of position that a correction is mandatory. At this point, the wind corrector along the Y axis brings the aircraft back to the position in the box from where it started the sequence, and the cycle is repeated.

Consider a crosswind that pushes the aircraft off centerline (the X axis) away from the judges. The aircraft starts the sequence on centerline and is gradually displaced away from centerline by the wind. Then comes the wind corrector along the Y axis, which brings the aircraft back to centerline, and the cycle starts again.

Now consider a crosswind that pushes the aircraft off centerline toward the judges. In this case, the aircraft starts the sequence parallel to centerline but upwind from it. By the fifth maneuver, it is pushed by the wind onto centerline. Now the wind corrector takes the aircraft upwind along the Y axis, away from centerline, back to where it started the sequence, and the cycle is repeated.

While wind correctors keep you within the box, they do little to help your presentation. Your aircraft presents an ever-changing perspective to the judges instead of remaining consistently nailed along centerline, increasing the risk of parallax error and making it more difficult to judge your performance.

One problem with wind correctors is inflexibility. What if there is no wind? You still have to fly the sequence, including the wind corrector. Now you have to play with positioning yourself along centerline in a no-wind condition to correct for the wind corrector! There are two types of wind correctors, high speed and low speed, to give you the advantage of the law of averages. The high-speed wind corrector is advantageous in a strong wind. The low-speed corrector is useful in light winds. You are at greatest disadvantage flying a low-speed wind corrector in a strong wind. You'll be motoring across the box for what will seem like hours. It will break the rhythm of your performance and will look BORING.

Another problem is that you have control only over inserting wind correctors in the freestyle sequences. What will you do if the designers of the known and unknown sequences don't insert wind correctors at points where you would like them to be or not at all?

At the end of the day, reliance on wind correctors to handle the wind is a good compromise technique. Aspiring advanced aerobatic pilots will find its use convenient as they gain experience. But beyond a certain skill level, the wind corrector becomes the lazy pilot's way of handling the wind. If you aspire to do well at the highest levels of national and international competition, the wind corrector technique will prove too restrictive. You have to learn other techniques that keep you firmly within the inner box in the worst allowable conditions and have the judges say, "Wind? What wind?" as they watch you fly and hold onto their hats.

Turning windward

The technique of turning windward is the most elegant solution, but it is very difficult to master because it requires constant control adjustments during the maneuvers. The idea is actually quite simple. With a coordinated control adjustment, you point the aircraft's nose into the wind, a few degrees off centerline. You fly the maneuver on this slightly offset longitudinal axis. Meanwhile, the wind pushes you back toward centerline. If you judged the offset angle right, the aircraft's path over the ground averages out to remain along centerline. Or, if you need to move up into the wind slightly, you make a greater adjustment throughout the maneuver.

What makes this technique difficult is that if the aircraft's attitude changes in a maneuver, causing the nose to begin to point away from the wind, you'll need to make an adjustment mid-maneuver to realign the nose into the wind. Let's consider the loop to explain this concept.

The aircraft is positioned along centerline with the wind blowing toward the judges. You are doing the loop from the judge's right to left; the wind is to your right. You pull into the loop, and as soon as you pull, you also roll slightly into the wind. The roll alone is enough to get the nose a few degrees into the wind. The aircraft is flying slightly away from centerline with its nose pointed away from the judges, while the wind acts to push the airplane back toward centerline.

At the top of the loop, you are slightly upwind of centerline. But now you have a problem. The nose is pointed away from the wind (you are inverted). If you just continued the loop, you would start flying downwind, and the wind would magnify the displacement. During the entire first half of the loop, you have to continuously turn upwind. At the top, you have to smoothly reverse the slight aileron input to keep continuously turning into the wind again during the second half of the loop. This technique will reestablish your flight path slightly upwind of the centerline, and by the time you complete the loop, you will be back on centerline, ready to position the nose upwind again for the next maneuver.

If you want to finish slightly upwind at the end of the loop, adjust the roll into the wind accordingly.

You can also make adjustments in other ways to other maneuvers. An example is the vertical roll. Assume, again, that the wind is blowing toward the judges, and you are pulling vertical from right to left. As you pull vertical, turn slightly into the wind. The wing (the aircraft's lateral axis) is now not precisely perpendicular to the centerline (X axis) because of the offset nose. The difference is not big enough for the judges to notice. However, if you were to do a vertical roll now and stop the wings perpendicular to centerline, the roll would appear to have been prematurely halted. You therefore need to keep rolling all the way to the point where you started the roll. The judges will see a perfect roll, not noticing the slight offset from centerline.

You can apply a wind-correction angle of as much as 10° if necessary before you look so out of position that the judges penalize you for it. At an airspeed of 200 mph, a 10° wind correction allows you to handle a veritable gale. In most situations, the wind-correction angle you need is much lower.

The big advantage of this technique is that the aircraft does not experience sideslip. The rudder is not loaded once the wind correction is made. The maneuvers can be flown as hard as in calm conditions. This method of wind correction is especially favored by French pilots and has been playing an important role in their impressive performance in recent international competition.

Sideslip

Another advanced wind-correction technique is the application of sideslip to remain aligned with the centerline (X axis). This technique is akin to staying on the runway centerline when landing by applying upwind aileron and opposite rudder. The aircraft is slipping in relation to the air mass but tracks true over the ground. This concept is easy to grasp but difficult to implement.

When you use this method of wind correction, you pull into the maneuver, apply aileron into the wind and opposite rudder to keep the nose aligned with centerline. Sideslip can be used to varying degrees depending on whether you want to maintain position over the ground or move upwind.

As we mentioned earlier, it might be surprising that sideslip may be used to correct for the wind, given our declarations in previous chapters that sideslip prevents the satisfactory completion of certain maneuvers, notably the vertical roll and tailslide. Our previous observations hold true, and this is the weakness of this wind-correction technique. There will be a sideways load component to the aircraft that makes vertical rolls and tailslides exceedingly difficult.

When flying maneuvers sensitive to sideslip, you have to apply sideslip on the entry, allow almost all of it to dissipate before you execute the maneuver, and reapply it upon completion of the maneuver. Sideslip won't dissipate entirely, and you might have to minutely correct for it as you fly the maneuver. For example, you will have to roll more slowly in the vertical roll and might have to chase the wing throughout the roll to keep the sighting device nailed to the horizon. The net effect, however, keeps you in close proximity of the centerline and makes your performance superior to the sequences of pilots who rely on wind correctors to handle the wind.

Phil Knight on conquering the wind

I have always believed that the actual performance of aerobatic figures should be a subconscious process. This allows the pilot to keep abreast of everything else that's going on during an aerobatic flight. I have come to the conclusion that the process of wind correction can and should also be a constant and subconscious exercise. Each time we fly, we focus on setting the aircraft up to be parallel to the X or Y axis. If we can accomplish this objective, then it should be no more difficult to set the aircraft up 5- to 10-degrees off the X or Y axis. If we choose to fly our flight over the centerline of the box, then each time we complete a figure and we are off the centerline, we input enough heading change to arrive at the other end of the box above this line.

This process, if practiced on each flight, becomes a subconscious reaction, just as did our ability to fly the aircraft parallel to the axis. This process should be used for up to 10° of directional change. In very strong winds that require corrections above 10°, the aircraft must be yawed so that it is not displaced from the axis to such a degree that it becomes easily visible to the judges. Yaw must be used with discretion, as its presence on vertical lines affects the presentation of the roll and can be a sure way to zero a tailslide. If a vertical line has no rolls on it, the aircraft can certainly be kept properly aligned by being slipped into the wind.

I think that these techniques need to be taught at all levels of aerobatics so that it becomes as natural to correct for the wind as it is to fly parallel to the axis of the box.

To this point, I've discussed wind on the Y axis (crosswind). Now let's consider wind on the X axis (headwind). We can control figure selection only in the freestyle program, but it is very important when flying high-performance airplanes to design a sequence that can handle all headwind velocities. Several rules I've come up with include not putting two hammerheads or a hammerhead and a humpty on the same line. I select figures for the downwind end of the flight that allow me to get away from that end as quickly as possible. Figures with only one vertical line, such as the shark's tooth, or with no vertical line, such as half loops or half Cubans, allow the time spent at the downwind end of the box to be minimal.

Whenever possible, always design humptys so that the radius over the top is into the wind. The known and unknown programs present a situation in which you must play the game with the situation you are dealt. A vertical line can be shortened by reducing power on the up line. This technique works well if there are no rolls on the line. Downlines can be shortened by exiting the line earlier. Both of these adjustments can affect the energy of the sequence and might require the pilot to start the flight higher than normal. These changes might also result in the degradation of the figures. Therefore, the pilot must weigh the cost of an out versus the loss of a grade on the figure to determine if this technique is appropriate.

Wind-correcting is very important to the success of a flight, both in penalty points and in the overall presentation of the flight. All competitors at all levels should make an effort to include wind-correction training in their training program.

Phil Knight is 1994 U.S. National Champion and a longtime member of the U.S. National Team. He currently competes in an Extra 300S.

Phil Knight

Phil Knight handling the wind.

15

Competition training

It is fairly easy to navigate your way through most advanced maneuvers, but to fly them well consistently, it is essential to engage in some form of structured training. A training regime needs to be tailored to your personal circumstances and objectives. The level and intensity of a personalized training program depends on, among other things, your level of aerobatic competence, aerobatic objectives, physical condition, mental ability to visualize your flights and handle stress, and the amount of time and money you are able and willing to spare for training.

In this chapter we discuss training from the competition pilot's point of view and present the ideal training regime for pilots preparing for a top-10 finish in the Unlimited U.S. nationals. For pilots with more modest goals, we offer options that are less strenuous and easier on the wallet. You'll also find this chapter useful if you don't fly competition but want to fly advanced maneuvers well in airshows or for your own amusement.

Shawn Tucker and Mike Goulian in a rarely seen aerobatic formation.

There is some confusion about what aerobatic competition training is. Many people think it is practicing your routine endlessly. Some think it is having an instructor in the back of the airplane offering helpful comments about your performance. Those who have some familiarity with competition flying know that much of competition training consists of a critiquer or coach observing you from the ground and evaluating your performance. In fact, competition training is all of these things and a lot more.

Make no mistake about it, aerobatics is an intensely physical sport, especially at the advanced level. Anyone who's experienced it knows that the body takes a tremendous pummeling on every advanced aerobatic sortie. Just like any athlete, an aerobatic pilot has to be in top physical shape to do well. Physical conditioning can only be attained by the right kind of strenuous exercise. A cunning plan to win the nationals by restricting yourself to light beer the night before the decisive flight won't do it.

Aerobatic flying, particularly advanced aerobatics, is also an intense mental process. You have to be able to visualize your sequences on the ground beforehand to be able to fly them effectively. There are a variety of visualization techniques that can greatly improve your aerobatic performance.

Competition flying is a mind game. It is of little use to have a top-gun fighter pilot's body if your mind blows a fuse when the pressure is on. Most top athletes in any sport work with sport psychologists to learn to handle the pressure to win and deal with the stress of a loss. Mental conditioning can also pay handsome dividends to the aerobatic competition pilot.

Let's take a closer look at the three main elements of an effective advanced aerobatic competition training regime: physical conditioning, mental training, and coaching.

PHYSICAL CONDITIONING

Physical conditioning is one of the most important aspects of aerobatic competition training, but it is given the least amount of attention by many pilots. Body weight, muscle strength and cardiovascular condition all affect how well your body can withstand the rigors of aerobatic flight. Consider two pilots who are both 5 feet, 9 inches tall. One weighs 160 pounds, the other 220 pounds. The 160-pound pilot will weigh 1600 pounds in a 10-G pull. A 220-pound pilot will weigh 2200 pounds in the same pull. The heavier pilot's heart, which already starts with the handicap of having to pump blood through several extra miles of blood vessels, now has to pump against an extra 800 pounds. That can get quite exhausting.

It is true that the overweight pilot has marginally higher G tolerance because his less-efficient circulation won't allow the blood to drain as rapidly toward the direction of the G load as does the trim pilot's cardiovascular system. That's nice when you are Sunday looping and can take a leisurely scenic tour to rest up between the slow roll and the Immelman. But its benefits are outweighed (literally) in an advanced aerobatic competition event when you have to perform as many as 16 advanced maneuvers in about six minutes. Halfway through the sequence you'll become utterly drained of energy because of your excess weight's extra strain on your heart, and you'll wish you had stayed on the Lazy Boy flying virtual snap rolls on the PC, sucking popcorn.

Being trim helps, but even the trim pilot can perform much better if he or she is in good physical shape. Strong and flexible muscles and a lot of energy dramatically improve endurance.

An essential aspect of aerobatic flying is coordination between visual clues and control movement. Aerobatic pilots can benefit greatly from doing hand-eye coordination exercises. There is no escaping it: Exercise is good for you!

It is best to custom design a physical training program to meet your personal needs. The most effective, though also the most expensive, way to do it is to work on an ongoing basis with a good personal trainer at your local health club. While that is not for everyone, at a minimum it is well worth the expense to buy an hour or two of a good trainer's time and design a regime that you can administer yourself. You should find a personal trainer who has a lot of comparative experience with a wide variety of sports, which helps in assessing training needs in any sport. Any physical training regime should include these goals:

- Improve cardiovascular capability
- Increase muscle flexibility
- Increase muscle strength
- Improve hand-eye coordination

Mike Goulian has been working with professional physical trainers for a long time to stay in top physical and mental condition. Integral to his current training program is a whole-body approach to conditioning known as the Burdenko method, which consists of exercises personally developed for Mike, taking into consideration the physical and mental stress experienced in the cockpit during Unlimited aerobatic sequences. Mike has two different weekly exercise routines, one for staying in shape during the competition season and one for the off season. Both are outlined below to give you a flavor of what can be involved in maintaining a competitive physical and mental condition.

Mike Goulian.

Conditioning program during the competition season

Monday: Burdenko land and water routine for flexibility and recovery

Tuesday: Weight room, basic full-body routine; Burdenko water routine for strength maintenance and recovery

Wednesday: Tennis or similar group activity

Thursday: Burdenko water routine for relaxation and recovery

Friday to Sunday: Off

Conditioning program during the off season

Monday: Exercise bike warm-up; weight room, traditional strength training

Tuesday: Burdenko water and land exercises, aerobic and anaerobic level

Wednesday: Burdenko land exercises to warm up; weight room, body routines, strength and cardiovascular endurance; water conditioning for flexibility, balance, and coordination

Thursday: Tennis, golf, or bike riding

Friday: Burdenko land exercises for warm up; weight room, traditional strength training; water coordination for flexibility, coordination, and recovery

Saturday: Off

Sunday: Tennis, two hours

If you believe that it is not whether you win or lose but how you play the game, you might not wish to go to the extremes of a top national competitor's physical-training program. You can indulge in a more moderate gym routine.

A good alternative to a moderate gym workout program might be playing a sport that combines all the suggested elements of the exercise regime. Squash, racquetball, basketball, and tennis (in that order) are all excellent training sports for aerobatics. They all involve cardiovascular, muscle stretching and strengthening, and hand-eye coordination components. Other excellent options are biking, rowing, rollerblading, and skating. Play any of these sports hard two or three times a week and do some supplementary weight training at home, and you should be in good shape for the rigors of most aerobatic competitions.

MENTAL TRAINING

The two important components to an aerobatic competitor's mental training are visualization and stress management. Visualization is a tool to help the pilot mentally concentrate on flying a sequence by visualizing it shortly before the flight. Stress management is being able to control the mental pressures that flying imposes on the pilot. In its basic form, it is learning to cope with the pressure of an imminent flight that is personally stressful for whatever reason. For competition pilots, it is learning to cope with the stress of having to fly in front of your peers.

Visualization

We've all seen them at aerobatic events: self-absorbed pilots silently walking around in some quiet corner away from the crowds, doing a strange routine that appears to combine elements of tai chi, karate, and ballet. They are visualizing by mentally flying through their upcoming sequence. This mental prerecording of the sequence is very effective in helping the pilot focus and precisely recall the maneuvers during the flight. It is especially useful for avoiding a split-second mental block at a crucial moment that earns a dreaded "zero" score for a maneuver and causes the pilot's standing in the competition to plummet.

Visualization was demonstrated to be a highly effective learning tool in an interesting experiment with basketball players. The players, all of equal competence, were divided into three groups. One group was trained to shoot baskets for two weeks, one group was trained for one day, and one group was trained for one day and told to visualize shooting baskets for the rest of the two weeks. At the end of the two weeks, each group took the same test shooting baskets. The group that shot baskets for two weeks did the best, but the group that trained for only one day and visualized shooting baskets for the rest of the time did almost as well. The group that trained for only a day trailed far behind.

All pilots can benefit from visualization throughout their initial and recurrent flight training. Aerobatic pilots find it invaluable throughout the never-ending learning process. The pilots we see walking through their upcoming flights at a competition have done it countless times before while learning and rehearsing the sequences. Visualization is an especially important tool for preparing to fly the unknown event. Since you are not permitted to rehearse the sequence, the only way you can prepare for it is by visualization.

There are no hard and fast rules for how to do visualization. Each pilot has to experiment and see what works best. Generally the idea is to imagine the flight as if you were sitting in the cockpit. Walk along the airplane's flight path, act out the airplane's attitude with your hands, visualize or even act out critical control movements and power adjustments, look where you would look sitting in the airplane, imagine what you would see, and imagine the engine sounds.

The modern aerobatic airplane cockpit.

Mike Goulian's precompetition visualization

A good way to gain an insight into what visualization is all about is to have Mike Goulian tell us in his own words how he visualizes a sequence just prior to takeoff:

I visualize my sequence about 45 minutes to half an hour before a flight. I listen to music on a headset while I'm doing it. The music is white noise that drowns out the surrounding environment and helps me concentrate. I visualize the sequence three times. I walk through it once, I visualize it sitting in the cockpit, and I visualize it from the judge's perspective.

First I walk through the sequence. I visualize the view as it would look from the cockpit looking outside. I look in the appropriate directions, I imagine the control inputs and the G forces, and I concentrate on the position of the wings at the appropriate points in the maneuvers. I do the whole sequence uninterrupted. If something breaks my concentration, I start from the beginning.

Next I sit in the cockpit and visualize the sequence a second time. The perspective is again from the cockpit looking out. The purpose of this visualization is to adjust the sequence for the local environment and prevailing conditions to most effectively present the flight to the judges. I concentrate on the control inputs required to adjust for the wind, the position of the aircraft in the box accounting for the wind, and the position of local landmarks at critical stages of the flight.

I visualize the flight for the third time sitting on the ground from the judges' perspective. This time I am looking up at my airplane, watching it fly the sequence, trying to spot any potential weaknesses in presentation, thinking of any additional adjustments required by prevailing conditions. I find it important not to be rushed through the visualizations and plan to complete them about 10 minutes before the flight. You don't want to cut it close, but you don't want to complete them too early either because then their effectiveness begins to diminish.

COACHING

Everyone thinks of coaching as the traditional form of aerobatic competition training. Since the term is often misunderstood, it needs to be clarified. Let's start with defining what a coach is. A coach is a person who evaluates your competition flying skills and is able to tell you exactly what to do to improve it and correct your mistakes. To do these tasks, a coach has to be a pilot who is an advanced aerobatic instructor and an experienced competition pilot.

This might sound obvious, but it isn't. Many people think that a judge can be an effective coach. A judge can be an effective critiquer, able to tell you what is imprecise about the appearance of your maneuvers. But unless he or she is also an experienced aerobatic instructor and competition pilot, a judge cannot tell you what to do differently in the cockpit to correct your errors.

Some people think a good aerobatic instructor can be a good coach, but unless that instructor is also an experienced aerobatic competitor, he or she is likely to have difficulty identifying the subtle errors in maneuvers that make a difference between a high score and a mediocre score.

What is coaching? It has two objectives:

- To train you to fly with precision
- To develop and improve your style

To meet these objectives, the coach spends a great deal of time watching you from the judge's perspective, evaluating your performance in the aerobatic box. You carefully review your flights together, devise solutions to any problems or weaknesses, and then go up and try the routine again. Typically you fly two sessions a day, each of about 15 to 18 minutes duration, and do a lot of work on the ground in between.

On the coached flights, you alternate flying full sequences with partial sequences that emphasize particular areas of your flying that need work. You train for the knowns by flying the known sequence, and you prepare for the unknowns with full or partial unknown sequences given to you by your coach. You and your coach are in radio contact, and each flight is also videotaped from the ground for postflight evaluation.

It is very important to develop a good rapport with your coach and to keep in mind that the coach is there to help you, not to unilaterally dictate to you. The relationship with a coach can get tricky if you already have an identifiable style that differs from the style the coach might prefer. The coach's job is to help you develop your style rather than chuck it and impose his or her style.

Coaching is a time-consuming, subtle process. There are no quick fixes that can be dispensed in one weekend seminar. When you hire a coach, you should plan to train for at least three days to realize a significant benefit. A week of training at a time followed by a lot of practice on your own seems to yield excellent results. At the national level, you should ideally receive one week of coaching per month throughout the year. At the regional level, try to arrange for a week of coaching three or four times a year.

There is also a role for dual instruction in a comprehensive coaching program, though it is only a supporting one. Dual instruction affords an opportunity for the coach to observe your handling of the controls firsthand and to help you polish your technique. One to three weeks' worth of dual per year (assuming two flights a day) are sufficient for the serious aerobatic competitor. Do bear in mind that this dual is distinct from dual instruction aimed at teaching you new maneuvers. Think of it more as a proficiency check.

Coaching is pricey, and, depending on your competition objectives and personal circumstances, you might decide that hiring a personal coach is not an option for you. You might consider instead several "budget" alternatives. The best option is to band together with three or four of your peers and hire a coach together for a week on a package deal. This practice is fairly common and has the added benefit of allowing the opportunity to learn from each other's performance. You can practically turn such a session into a tutored personal mini-contest. A number of top coaches run structured training camps for groups of pilots, offering competition training in a marked box for all competition categories.

If sharing a coach doesn't prove feasible, another option is to get together with another competition pilot who is as experienced as you are and coach each other. You can see each other's errors, and since you are also both experienced pilots, you can experiment with solutions. It is less effective than getting an experienced coach, but it has benefits. It helps if you both have judging experience.

As a last resort, you can also get a judge to critique you, but then you are on your own to determine the causes of any errors and devise solutions. Don't get us wrong. Critiquing is a very useful periodic reality check and well worth arranging, but don't consider it a substitute for coaching.

Coaching by an experienced coach is the best possible thing you can do to improve your aerobatic flying. You can only do so much practice on your own to increase your competence. Eventually you will just be making the same subtle errors over and over again without realizing it. Your efforts can even become counterproductive, because you will be reinforcing bad habits, making it more difficult to break them. A good coach can show you in one week ways to improve your flying that you would never discover on your own and can propel your performance to a whole new level. Your first week of working with a top coach can be a real revelation.

Steve Victorson on Mike Goulian's physical fitness program

Physical conditioning for aerobatic pilots can take a variety of forms and depends on the pilot's physical condition and the level of intensity with which the pilot performs aerobatics. Any program should address physical as well as mental strength. Perhaps the best way to give you a flavor of what is involved in keeping a top aerobatic pilot in shape is to review Mike Goulian's physical conditioning program, described in Chapter 15.

Mike Goulian's program is designed to enhance his general level of physical fitness as well as to improve his flying-specific physical conditioning. His program consists of standard machine and free weight strength-training exercises, the "Burdenko Method," which is a whole-body exercise program I'll describe momentarily, and a variety of recreational activities such as tennis, golf, bike riding, and skiing. The goal is to develop Mike into a highly conditioned athlete capable of withstanding the physical and mental demands of Unlimited aerobatic training and competition.

In the weight room, Mike has routines that utilize machines and free weights and routines that involve moving the body with minimal added weight, such as push ups, dips, pull ups, and step ups with small hand weights. In general, the weight-lifting routines follow the basic principles of strength training, but Mike also does some variations designed specifically for flying.

For example, instead of doing two or three sets of 10 repetitions of a particular exercise, Mike does only one set of 10 repetitions, then moves, without rest, to another exercise and immediately starts the next set. His routine might consist of four to six different exercises and 40 to 60 total repetitions performed without rest. The goal is to develop strength endurance in more than one muscle group while simultaneously adapting physically and mentally to different movement patterns, similar to how Mike moves from one figure to the next when flying an aerobatic sequence.

An important consideration in designing such a routine is safety from physical risk. While aerobatics is physically and mentally draining, it does not require major body movement, so Mike's routine need not include risky Olympic lift routines of questionable safety. The sets of exercise routines should consist of familiar exercises and other exercises that require effort and concentration but have minimal risk of physical injury.

Another very important aspect of Mike's training program is the Burdenko Method, a whole-body approach to physical conditioning that forms the philosophical basis of Mike's entire conditioning program. Named after the Russian Igor Burdenko, who spent more than 30 years developing it, the method has been successfully used by pilots, astronauts, and numerous athletes from a variety of sports.

The Burdenko Method is based on functional exercises using the body in a functional position. Mike, for example, has routines designed in sitting, standing, and horizontal positions. It involves movement in every direction (forward, backward, up, down, and lateral), it develops both the body and

Steve Victorson

mind, and it utilizes both land and water. Thus, after every land session, Mike uses the pool for additional conditioning and/or relaxation and recovery.

A typical Burdenko land training session involves balance, coordination, strength, endurance, and flexibility conditioning, utilizing weight machines, small hand weights, sticks, or just the athlete's body. One routine Mike does in the Burdenko program is what we simply call the four-minute routine, during which Mike does a variety of exercises requiring immense physical and mental concentration. By design, the routine is equal in length to some of the most strenuous aerobatic sequences. The water is used in similar fashion, with flotation belts and kickboards replacing the weights.

You need not necessarily rush out and sign up for the Burdenko Method to become U.S. National Champion. It is only one of several methods that work well for aerobatic pilots. More importantly, remember that to reach peak aerobatic performance, you must be able to concentrate 100 percent on your flying. It is equally important for any exercise program to develop your ability to concentrate 100 percent as it is to get you in physical shape.

Steve Victorson, M.S., U.S. Ski Team conditioning coach from 1987 to 1991, is currently completing a doctorate in human movement at Boston University. He is also a private coach and trainer associated with the Burdenko Water and Sports Therapy Center in Lexington, Massachusetts.

16

Advanced aerobatic aircraft

A large number of aerobatic aircraft are available to the advanced aerobatic pilot. Some have been around for decades. Others have only recently appeared on the scene. They are made in quite a number of countries around the world, chief among them Germany, Russia, France, and the United States. Some of the older ones are from countries that no longer exist, such as Czechoslovakia. Some are products of major aircraft manufacturers, but many come from small cottage shops specializing in their production.

No listing of aerobatic aircraft can be all-encompassing, if only for the fact that by the time it appears in print, some new design is likely to be making its debut. In this chapter we present some of the most popular advanced aerobatic aircraft on the competition circuit today by manufacturer in alphabetical order. Many are in production and can be purchased new. Others are available in kit form, and many of them are available on the used market, which you are strongly encouraged to investigate. For brief descriptions of the older, lower-performance aircraft types not covered here, see *Basic Aerobatics*. For comprehensive technical descriptions of any aircraft, the best source remains *Jane's All the World's Aircraft*.

AKROTECH AVIATION

Richard Giles of Scappoose, Oregon, created the experimental kit-built Giles 200 and 202 to bring modern unlimited aerobatic performance to pilots who can't afford the 300-plus-hp super-expensive super-monoplanes. One of those pilots was Giles himself, and his design philosophy for the G-200 was simple. He didn't have marketing in mind when he sat down to dream up the G-200. "I designed this airplane for myself," he said. Perhaps that is the best way to ensure that what you'll create will be good enough to get the market to come to you. That, at any rate, is what happened to the G-200 when it made its debut at Sun 'N' Fun in 1994. Demand for kits was brisk, and there was such interest in a two-seater of comparable characteristics that Giles developed the G-202.

The Giles designs are, in essence, a return to the idea of a light, small airframe that can achieve, on a less powerful engine, performance similar to that of the larger, more powerful aircraft. Richard Giles achieved this objective by using light, high-strength composites, particularly prepreg carbon fiber for the airframe, and airfoil technology similar to what has proved so successful on the CAPs and Extras.

The Giles designs are truly contemporary in the sense that they are entirely composite, and the fuselage is of monocoque construction. This approach differs from the traditional opinion, which holds that wood, steel, and fabric are still the best option for homebuilts because they are easy to work with.

Here is what Giles has to say on the issue:

The question about structure can't be simply answered by just addressing the properties of carbon/epoxy composite or the properties of wood or the properties of metal. We have to look at the ability of the airframe to carry and transfer the loads of flight during aerobatic maneuvers.

When I first designed the G-200, I planned to build it with a steel tube and fabric fuselage and plywood-skinned wings, much like the Stephens Akro and the Pitts. But as I began to do the structural analysis on the airframe, it became clear that there were more efficient types of structures available than truss and fabric. Then, as I looked at the idea of a monocoque fuselage, I also began to look at materials that were more efficient for that type of design than wood, steel, or aluminum.

The G-200 and G-202 are, by definition of the design, available only in kit form with ready-made major molded components, making difficulty of composite construction for the home builder a non-issue. Giles has made arrangements with a variety of suppliers to manufacture the kit components. The kit has various options, including fast-build kits. Pilots who seek unlimited aerobatic performance in a modern, economical monoplane and enjoys homebuilding might find a Giles design an ideal choice. Estimated build time is 1000 to 2000 hours, depending on builder experience. If you are not a builder but want a G-200 or G-202, contact AkroTech to find out what options are available for obtaining a completed aircraft.

Giles G-200

The G-200 is a composite airplane. The wings, including the ribs and spar, the tail section, and the fuselage are all carbon fiber. The skins have a honeycomb core. Parts are laid up in female molds, vacuum bagged, and oven-cured at high temperatures. Spars and reinforcements are solid layups of carbon fiber and epoxy. The hinges and fittings are machined aluminum. Where the aluminum fittings or steel bolts come in contact with the carbon fiber, a stainless steel or fiberglass barrier prevents corrosion.

The G-200 is designed for Lycoming engines ranging from 150 hp to 220 hp and can be equipped with fixed-pitch or constant-speed propellers. Bear in mind how critical added weight is at the weight level of an airplane as tiny as the G-200. The added weight that is the cost of the extra performance reduces the net performance gain from the added power and accessories. The optimal combination must be carefully assessed prior to making the final decisions about engine, propeller, and accessory choices. One point to keep in mind is that a constant-speed propeller makes life much simpler in spite of the extra weight and cost.

The flying characteristics of the G-200 are said to be outstanding by all the competition pilots who have flown it. Especially memorable is the diminutive aircraft's roll rate, said to be in excess of 400° per second. An interesting characteristic, given the airplane's lighter wing loading (in comparison to the super-monoplanes) but high power loading, is that it can achieve the same maneuvers from lower entry speeds, resulting in lower Gs, greater ease in staying within the confines of the box, and less wear and tear on the pilot.

Visibility in the air as well as on the ground is outstanding. The pilot's seat reclines 45°, optimizing G endurance.

Whether this category of aircraft can compete successfully with the super-monoplanes remains to be seen, but it certainly provides an economical entry into modern unlimited competition for pilots with homebuilding skills.

Giles G-202

The G-202 is the two-seat sibling of the G-200, developed in response to popular demand. Although it looks similar to the G-200, it is a separate design. Its wingspan is larger by 2 feet and its fuselage is longer. Like the G-200, it has near full-span ailerons, giving it a roll rate said to be around 400° per second. The G-202 is designed for Lycoming engines between 150 and 230 hp. A 200-hp Lycoming IO-360 with a constant-speed propeller is said to be the optimal combination. The design principles, materials, and construction processes used in the G-202 are exactly the same as the G-200.

Jeffrey Zeckendorf

Giles 202

	G-200	**G-202**
SPECIFICATIONS		
Engine	Lycoming 320–360	Lycoming 320–360
Horsepower	150–220 hp	150–230 hp
Propeller	Various	Various
Wingspan	20 ft	20 ft
Wing area	75 sq ft	90 sq ft
Length	18 ft	20 ft
Height	5 ft 4 in.	5 ft 7 in.
Seats	1	2
Fuel capacity	18–48 gal	18–58 gal
WEIGHTS AND LOADING		
Gross weight	1300 lb	1600/1400 lb
Empty weight	750 lb	950 lb
Wing loading	17.3 lb/sq ft	13.3 lb/sq ft
Power loading	5 lb/hp	6 lb/hp
Limit load (pos/neg)	± 10 G	± 10 G
PERFORMANCE		
Vne	220 kt	220 kt
Cruise speed	185 kt	176 kt
Rate of climb, sea level	3500 ft/min	3500 ft/min
Roll rate	400 + °/sec	400 + °/sec
Stall speed	57 kt	51 kt

AVIAT

The Pitts Special is an icon of American aerobatics that had its beginnings as a homebuilt. It went into production in single-seat and two-seat certified versions in 1971 in Afton, Wyoming, in a joint venture between Curtis Pitts and Herb Anderson. Eventually the factory was bought by Frank Christensen. Known as Christen Industries, it produced the Pitts as well as kits of another great aerobatic two-seater developed from the Pitts by Christensen, the Christen Eagle. When Christensen sold the company, it became AVIAT, a name retained through subsequent changes of ownership. Today AVIAT continues to produce the-two seat Pitts S-2B and the single-seat S-1T and S-2S. The Eagle kit is also still available by special order.

As we go to press, AVIAT is developing yet another single-seat Pitts, the S1-11B. Equipped with a 300-hp Lycoming and a larger redesigned airframe, it promises to outperform its predecessors and might give the 300-hp monoplanes a run for their money. Another purpose for developing the airplane is to use it as a test bed for a possible two-seat version intended to be a step up from the S-2B.

For a historic retrospective on the Pitts Special, see page 67.

Pitts S-1S

Making its appearance in the mid 1960s, the Pitts S-1S was the first Pitts Special fully capable of modern unlimited aerobatics. It was developed from the Pitts S-1C, and like all Pitts Specials, it had a fabric-and-metal-covered steel tube fuselage, a fabric-covered wood wing, pushrod-controlled ailerons and elevator, and cable-controlled rudder.

The major change was a new pair of wings that was more symmetric than the S-1C's wing, hence it was capable of high-performance inverted flight. Designing the wings was a challenge because of the biplane configuration. The top wing of the original Pitts was set to stall before the bottom wing to provide good stall characteristics. In inverted flight, this arrangement caused an obvious problem. Pitts experimented with symmetrical upper and lower airfoils as early as 1960, but he scared himself so much in the stall that he began dismantling the airplane as soon as he landed. He then tinkered with the airfoil design and came up with a proprietary solution that solved the problem and gave the airplane superb inverted handling. For many years Pitts did not make the plans available for this wing. He sold completed wings to homebuilders and produced kits and the certified version of the S-1S. The factory finally made the wing plans available in the late 1970s.

In addition to the new wing, the S-1S also had a 180-hp Lycoming O-360 instead of the 125-hp Lycoming and upper and lower sets of ailerons (an electrical system was optional). These changes yielded a dramatic increase in vertical penetration and roll rate. In 1972, American pilots flying the S1-S won both the men's and women's world aerobatic championships, and even today the aircraft is a formidable competitor in all but the unlimited category.

Pitts S-1T

The S-1T was an upgrade of the S-1S with a 200-hp fuel-injected Lycoming IO-360, a Hartzell constant-speed propeller, and a standard electrical system. It became available in 1981. The increase in weight that came with the engine and system enhancements offset much of the expected performance enhancements but made the airplane more convenient to operate than the S-1S.

Pitts S-2A

The S-2A was the first two-seat Pitts and the first Pitts to be certified and put in production. Introduced in 1971, it was a godsend for aspiring advanced aerobatic pilots in search of dual instruction in an aircraft capable of advanced aerobatics. While similar in appearance to the S-1 series aircraft, the bigger S-2A is an entirely separate design. It is equipped with a 200-hp fuel-injected IO-360 Lycoming engine, a Hartzell constant-speed propeller, and a standard electrical system. The S-2A was in production until 1982.

Curtis Pitts' S-2 prototype had two open cockpits. Early production models had the small bubble canopy option for the rear seat, and most were made with the big greenhouse bubble canopy that enclosed both seats.

Pitts S-2B

Introduced in 1983, the S-2B was an upgrade of the S-2A, resulting in a remarkable increase in performance that made available a two-seat Pitts that was competitive at the unlimited level. Power was increased to 260 hp supplied by a Lycoming IO-540. Other than the structural modifications required to accommodate the bigger engine, the S-2B is essentially the same airplane as the late-model S-2A. It continues to be one of the most popular two seat advanced aerobatic aircraft and remains in production.

Pitts S-2S

The Pitts S-2S is essentially a single-seat S-2B, although it actually preceded the S-2B by a couple of years, serving as the test-bed aircraft for reequipping the S-2A with the bigger 260-hp engine. In the S-2S, the front seat is replaced with a 35-gallon fuel tank, giving the airplane a range of over 400 nautical miles.

The S-2S might have been the highest-performance single-seat Pitts, but by the time it appeared, the monoplanes were beginning to rule the highest levels of unlimited competition. Most potential buyers in the market for S-2S-level performance opted instead for the extra seat of the practically similar-performing S-2B. More than 30 S-2S models were built to special order and are still available from AVIAT today.

Eagle

The Eagle is arguably the most beautifully made, most comprehensive kit airplane made to this day. The meticulous craftsmanship, packaging, and building instructions have become almost as much a legend on the kitplane scene as is the Pitts on the aerobatic scene.

The Eagle came into being when Frank Christensen couldn't quite come to terms in acquiring the Pitts line and decided to go ahead and develop a kitplane based on the S-2 series Pitts, incorporating his own ideas for improvement and for making it a good kitplane from the builder's perspective.

An issue with the S-2 series for pilots who want more than just an aerobatic airplane has been that they are essentially utilitarian aircraft intended solely for the purpose at hand. If you want to fly unlimited aerobatics but would also like to use the airplane as a weekend runabout, the S-2's cockpit is just a bit too cramped, its panel just a bit too spartan. In addition, many pilots voiced the opinion that somewhat tamer ground handling would also be nice. Frank Christensen addressed all these issues in creating the Christen Eagle. Ironically, a few years later, he was also successful at acquiring the Pitts line.

The Eagle is basically a refined S-2A. The airfoil and engine are the same, but the Eagle has less dihedral in the lower wing, its center of gravity is slightly further aft, its cowling is more streamlined, its rudder is slightly larger, and the wheelbase of its spring-leaf landing gear is wider. The Eagle's bubble canopy comes further down on the sides of the cockpit, slightly improving visibility, and the cockpit is wider and more roomy than the Pitts.

	Pitts S-2S	Pitts S-1T	Pitts S-2B	Eagle
SPECIFICATIONS				
Engine	Lycoming AEIO-540	Lycoming AEIO-360	Lycoming AEIO-540	Lycoming AEIO-360
Horsepower	260	200	260	200
Propeller	constant speed	constant speed	constant speed	constant speed
Wingspan	20 ft	17 ft 4 in.	20 ft	19 ft 11 in.
Wing area	125 sq ft	85 sq ft	125 sq ft	125 sq ft
Length	17 ft 4 in.	15 ft 6 in.	18 ft 9 in.	18 ft 6 in.
Height	6 ft 7 in.	6 ft 3 in.	6 ft 7 in.	6 ft 7 in.
Seats	1	1	2	2
Fuel capacity	35 gal	20 gal	29 gal	25 gal
WEIGHTS AND LOADING				
Gross weight	1500 lb	1150 lb	1625 lb	1525 lb
Empty weight	1100 lb	830 lb	1150 lb	1050 lb
Wing loading	12 lb/sq ft	11.7 lb/sq ft	13 lb/sq ft	12 lb/sq ft
Power loading	5.77 lb/hp	5.75 lb/hp	6.25 lb/hp	7.6 lb/hp
Limit load (pos/neg)	+6/–3	+6/–3	+6/–3	+6/–3
PERFORMANCE				
Vne	176 kt	176 kt	185 kt	175 kt
Maximum speed, sea level	163 kt	160 kt	162 kt	160 kt
Rate of climb, sea level	2800 ft/min	2800 ft/min	2700 ft/min	2120 ft/min
Roll rate	240°/sec	270°/sec	240°/sec	185°/sec
Range	352 nm	268 nm	277 nm	305 nm
Stall speed	50 kt	56 kt	55 kt	50 kt

AVIONS MUDRY

Avions Mudry was founded in 1953 as a light aircraft and glider manufacturer. Between 1965 and 1970 the company developed the CAP 10B aerobatic side-by-side two-seater. The wood-and-fabric airplane was a great success. It was capable of advanced aerobatics but could also be used as a training/touring aircraft. It became a primary trainer for the French air force and navy and also became a popular club aircraft, primarily in Europe.

The CAP 10B's success solidly established Avions Mudry as an aerobatic aircraft manufacturer. Its first competitive single-seater was the CAP 20, introduced in the late 1960s. It made its first appearance in world competition at the 1970 World Aerobatic Championships in Hullavington, England. It was basically a modified CAP 10B. It had the same airfoil and wing plan but no flaps and was structurally stronger. It was equipped with a 200-hp Lycoming engine and a Hartzell constant-speed propeller. The spar was a special kind of oak rather than spruce, and the airframe was a bit on the heavy side. Although a fine unlimited aircraft, the CAP 20 proved to be less agile than the Pitts and wasn't sufficiently competitive.

The next model was the CAP 20L. It had the same engine and airfoil as the CAP 20, but the wing was shorter and had less dihedral (resulting in an improved roll rate), and it had a lighter fuselage.

The CAP 20L was followed in 1980 by the CAP 21, which had the same engine and fuselage as the CAP 20L but a new wing with a better airfoil and a further improved roll rate. It was, however, unable to keep up with the 260-hp and 300-hp monoplanes then coming on line, and it was succeeded in 1985 by the CAP 230, a totally new airplane.

During the late 1980s, the CAP 230 was refined into the CAP 231, and in the hands of a succession of outstanding pilots, this airplane reaped a lion's share of world aerobatic championship medals during the first half of the 1990s. The CAP 231 was followed in 1995 by an even lighter and aerodynamically more-refined model, the CAP 232.

Mudry's big contribution to aerobatics was the pioneering use, starting with the CAP 21, of a special airfoil designed to maximize airflow attachment during radical maneuvering, which has set the standard for the airfoils of modern aerobatic aircraft.

CAP 10B

For pilots who want an airplane capable of flying advanced aerobatics and doubling as a fast, economical VFR cross-country touring machine, the best choice might well be the CAP 10B. Although the design is several decades old, this perky, bubble-canopied side-by-side two-seat monoplane continues to be one of the most outstanding all-around sport aircraft in the world. It is of wood and fabric construction and is powered by a 180-hp Lycoming with a two-blade, fixed-pitch, composite Hoffman propeller.

CAP 231

CAP 231/231EX

Based on Mudry's experience with the CAP 21, the CAP 231 was an entirely new airplane designed to compete with the German and Russian aircraft that were moving to the forefront of unlimited aerobatic competition in the late-1980s. A carefully selected airfoil, good weight control, and a 300-hp engine made the wood-and-fabric CAP 231 a top contender in unlimited events. Its only drawback, a relatively slow roll rate of 270° per second, proved no handicap, for the world's best pilots flew it to top honors at the 1990 World Championships at Yverdon, Switzerland. Flying CAP 231s, Claude Bessier and Patrick Paris of France took gold and silver overall in the men's event, while Linda Meyers of the United States won silver overall in the women's event.

The CAP 231EX is essentially a CAP 231 with a carbon-fiber wing built by Walter Extra. The market demanded carbon-fiber wings on the CAP 231, and Mudry responded through the cooperative arrangement with Extra. The wing was essentially the same wing as the one on the Extra 260, which was modified for mating to the 231 fuselage. The new wing, with a larger wing area and ailerons along 75 percent of the span, increased the roll rate to 330° per second. The cockpit interior was also entirely redesigned for increased pilot comfort.

The CAP 231EX was flown by the French team at the 1994 World Aerobatic Championships in Budapest, Hungary, where two of its members, Xavier de L'Apparent and Christine Genin, won the overall titles and the French team took the team title. Mudry and Extra chose to discontinue their association after the introduction of the EX model, which presented Mudry with the task of constructing its own wing for the series and led to the development of the CAP 232.

CAP 232

The CAP 232's most important feature is a brand-new wing design, based on Mudry's experience with the CAP 231 and 231EX. The 232's wing is made of high-performance, preimpregnated carbon fiber and structural resin, cured at high temperature under high pressure in an autoclave. The new airfoil provides increased strength and rigidity and lower weight. As a result, the 232's empty weight is less than 1300 pounds. According to Mudry, the lower inertia resulting from the lower weight improves the 232's roll rate in comparison to the 231's to 420° per second.

	CAP 10	CAP 231EX	CAP 232
SPECIFICATIONS			
Engine	Lycoming AEIO-360	Lycoming AEIO-540	Lycoming AEIO-540
Horsepower	180	300	300
Propeller	Hoffman	MTV	MTV
Wingspan	26 ft 5 in.	24.3 ft	24.3 ft
Wing Area	116.8 sq ft	109.2 sq ft	109.2 sq ft
Length	23 ft 5 in.	22.2 ft	22.2 ft
Seats	2	1	1
Fuel capacity		55 gal	55 gal
WEIGHTS AND LOADING			
Gross weight	1830/1675 lb	1810/1610 lb	1810/1610 lb
Empty weight	1210 lb	1330 lb	1290 lb
Wing loading (aerobatic)	14.3 lb/sq ft	14.7 lb/sq ft	14.7 lb/sq ft
Power loading (aerobatic)	9.3 lb/hp	5.4 lb/hp	5.4 lb/hp
Limit load (pos/neg)	+6/−4.5	+10/−10	+10/−10

	CAP 10	CAP 231EX	CAP 232
PERFORMANCE			
Vne	184 kt	219 kt	219 kt
Maximum cruise speed	146 kt	189 kt	189 kt
Rate of climb, sea level	1600 ft/min	3400 ft/min	3400 ft/min
Roll rate	180°/sec	330°/sec	420°/sec
Range	540	650	650
Stall speed	51/43 kt	54 kt	56 kt

EXTRA FLUGZEUGBAU

Walter Extra, a German homebuilder with several Pitts to his credit, saw an opportunity to commercialize the Laser design in the early 1980s. He developed a derivative design inspired by the Laser, which he called the Extra 230. The Extra 230, powered by a four-cylinder 230-hp Lycoming, looked like a Laser and used similar steel-tube, wood, and fabric construction techniques, but it was different in many important respects.

A key feature was a new airfoil, similar in concept to the one used on the CAP 21 but developed by German aerodynamicists during World War II. The most notable visual trait of the airfoil is a thick, square trailing edge that doesn't suggest aerodynamic smoothness to the uninitiated but, in fact, has highly desirable characteristics for aerobatic aircraft. The airflow coming off the bottom of the trailing edge creates an eddy, which results in a low-pressure area immediately aft of the trailing edge. This low-pressure area pulls the airflow off the top of the wing at a higher speed, resulting in increased lift at a particular airspeed. Another advantage of the airfoil is that during rapid changes in angle of attack, the airflow separates from the wing surface at a later stage than on most other airfoils previously used on aerobatic aircraft, enabling the Extra to corner particularly well.

The 230's wing was also constructed differently from the Laser's. It had a box spar instead of an I-beam and was made of stronger Polish pine. There were no lightening holes in the wing ribs. These features made the wing stronger than the Laser's wings.

The fuselage was made of heftier steel tubing. An aluminum turtleback could be removed in minutes, revealing the entire inside of the aft fuselage, a feature that has become standard on all subsequent Extra aircraft.

The Extra 230 had a lower stall speed than the Pitts, it accelerated more rapidly and cornered better, and its clean, straight lines presented the aircraft better to the judges. Clint McHenry was an early convert to the Extra 230 and flew it to several U.S. national championships as well as to high placings in the world championships.

The Extra 230 was an instant hit, and its success led to the development of the Extra 260, the first Extra with carbon-fiber composite wings. In 1991, flying an Extra 260, Patty Wagstaff became the first woman ever to win the U.S. nationals.

The Extra 260 was followed by the two-seat, 300-hp Extra 300 and the single-seat Extra 300S. Notable on the 300S was the relocation of the wing from the mid-wing to the low-wing configuration, greatly improving visibility. The 300 was the first modern unlimited aerobatic two-seat monoplane to be certified. The 300S also received its certification at the same time, and from this point on all Extra aerobatic aircraft have been certified aircraft.

The 300S became the most popular unlimited aerobatic aircraft in the world and inspired a two-seat replacement for the 300, which was designated the 300L.

For aspiring advanced aerobatic pilots and aerobatic schools, perhaps the most exciting recent development at the Extra factory is the introduction of the certified Extra 200, a 200-hp two-seater designed to be a tamer, more economical version of the super-monoplane.

Based on the quality and performance of its airplanes, the number of aircraft sold, and the depth of its product line, during the 1990s Extra has become the world's leading producer of high-performance aerobatic aircraft.

Extra 300

The Extra 300 was the first two-seat super-monoplane to be certified. Certification made the 300 eligible to be used as an advanced aerobatic training aircraft by flight schools. It has carbon-fiber composite wings, tail, and controls, and a steel-tube, fabric, and aluminum fuselage. The Extra 300 is powered by a 300-hp Lycoming AEIO 540 engine and is equipped with a three- or four-blade, composite, constant-speed MTV propeller.

Extra 300S

The Extra 300S became Extra's most successful model shortly after it was introduced. The ultimate single-seat competition machine, it is derived from the Extra 300. It utilizes the same wing, engine-propeller combination, and construction techniques. Its most important innovation in comparison to the Extra 300 was the relocation of the wing to the low-wing configuration. This change not only greatly improved visibility on takeoff and landing, but also made the 300S one of the most aesthetically pleasing aircraft of its kind.

Extra 300L

The Extra 300L was the logical development of the Extra 300S. Following the success of the low-wing, single-seat model, Extra introduced the 300L as a two-seat version of the 300S and a replacement of the 300. There is very little difference between the 300L and the 300S except for the extra seat.

Extra 200

The Extra 200 was introduced to meet a very obvious need for a two-seat certified unlimited aerobatic monoplane that was less demanding than the 300L and more economical to operate. Essentially, the 200 is a 300S with a 200-hp Lycoming AEIO-360 engine and a dual cockpit. The wings and tail are identical to the 300S. A two-blade, composite, constant-speed MTV propeller is standard, and a three-blade propeller is optional.

This aircraft is the only certified aircraft of its kind and should prove extremely popular with flying schools. Costing about two-thirds of the price of a 300L and much more economical to operate, it should become the standard two-seat advanced-aerobatic trainer for many years to come. For pilots who prefer a single-seat environment, the 200 has a "single-seat" bubble canopy structure designed to cover up the front seat, which can be interchanged with the two-seat canopy in minutes.

	Extra 200	Extra 300	Extra 300L	Extra 300S
SPECIFICATIONS				
Engine	Lycoming AEIO-360	Lycoming AEIO-540	Lycoming AEIO-540	Lycoming AEIO-540
Horsepower	200	300	300	300
Propeller	MTV	MTV-9	MTV-9	MYV-9
Wingspan	24.6 ft	26.3 ft	26.3 ft	26.3 ft
Wing Area	112 sq ft	115 sq ft	115 sq ft	112 sq ft
Length	22.3 ft	23.4 ft	22.8 ft	21.8 ft
Height	8.4 ft	8.6 ft	8.6 ft	8.6 ft
Seats	2	2	2	1
Fuel Capacity	30/40 gal	42 gal	44 gal	45 gal
WEIGHTS AND LOADING				
Gross weight	1914/1848 lb	2095/1808 lb	2095/1918 lb	2028/1808 lb
Empty weight	1199 lb	1470 lb	1474 lb	1343 lb
Wing loading (aerobatic)	16.5 lb/sq ft	16.6 lb/ sq ft	16.6 lb/sq ft	16.1 lb/sq ft
Power loading (aerobatic)	9.2 lb/hp	6.0 lb/hp	6.4 lb/hp	6.0 lb/hp
Limit load (pos/neg)	+10/–10	+10/–10	+10/–10	+10/–10
PERFORMANCE				
Vne	220 kt	220 kt	220 kt	220 kt
Maximum speed, sea level	158 kt	178 kt	180 kt	185 kt
Rate of climb, sea level	1600 ft/min	3200 ft/min	3200 ft/min	3200 ft/min
Roll rate	340°/sec	340°/sec	360°/sec	380°/sec
Range	450/600 nm	415 nm	425 nm	469 nm
Stall speed	56 kt	57 kt	57 kt	55 kt
Takeoff ground roll		320 ft	305 ft	315 ft
Landing ground roll		565 ft	538 ft	561 ft

ONE DESIGN

The IAC One Design was the result of an effort to develop a standard aircraft that would be flown in its own class to level the playing field and make unlimited aerobatics available at a more affordable cost. The one-design idea is borrowed from the world of competitive sailing, where such classification is common practice. Its design objective is to provide an inexpensive airframe combined with a four-cylinder engine in the 150- to 200-hp range that will perform unlimited aerobatics at a competitive level.

The One Design is intended to be a homebuilt airplane. Arrangements have been made for Aircraft Spruce and Specialties to supply the plans and all the required material in a complete package. Several specialty builders are also in the business of providing prefab components for builders who want to save construction time.

The One Design, developed by Dan Rhin, is of traditional construction. The fuselage is fabric-covered steel tube, the wings are wood reinforced with a composite layer. This type of low-tech construction is ideal for homebuilding, especially for builders interested in building from scratch. It allows for easy custom fabrication of components from raw stock, provides a structure as strong as other, less-forgiving construction techniques, and has a well-established track record in the homebuilding community.

The One Design is a contemporary of the Giles 200, with similar performance. The two aircraft appeared at about the same time, and many aerobatic enthusiasts fondly remember their friendly aerial duels at Oshkosh and Sun 'N' Fun.

Whether or not the one-design competition concept will be successful can only be determined when there are sufficient One Designs built to hold competitions. As this book was going into print, Dan Rhin had also just completed a two-seat version of the airplane.

SPECIFICATIONS	DR-109 (two seater)
Engine	Lycoming
Horsepower	200/260/300
Propeller	Various
Wingspan	24 ft
Wing area	114 sq ft
Length	22 feet
Height	—
Seats	2
Fuel capacity	45 gal

WEIGHTS AND LOADING	DR-109
Empty weight	1260/1350/1420
Limit load (pos/neg)	+/–10

PERFORMANCE	
Maximum speed, sea level	195/205 mph
Roll rate	360+°/sec
Stall speed	59/62 mph

STAUDACHER

For many years Jon Staudacher of Kawkawlin, Michigan, was America's premiere builder of hydroplane racing boats. In his free time, he also built airplanes for his own amusement, among them a Pitts Special. One day he broke his leg and was laid up for several months. He used the time to design a world-class aerobatic monoplane, the Staudacher S300, which led him to a new career designing and custom-building unlimited aerobatic aircraft.

There are two Staudacher models: the single-seat 300 series and the two-seat 600. No two Staudachers are exactly the same since custom-building each aircraft to special order allows Jon to make changes and improvements to each airplane he builds.

Staudacher relies mostly on traditional construction methods that have been in use since the 1930s. Tube, fabric, and wood construction is labor-intensive but requires little capital, so it is ideally suited to Staudacher's small, custom production runs. "High-tech composite construction isn't for us," said Jon, "because even though it is less labor-intensive, it demands immense amounts of capital for the molds and autoclaves and requires mass production to be cost-effective."

An added benefit to traditional methods is greater ease of making modifications on the production line. "In large measure, aerobatic aircraft design is still black magic," said Staudacher. "You can't simulate an entire aerobatic sequence in the wind tunnel, and we often try changes from one production aircraft to the next. That would be practically impossible if you are locked into the constraints of composite production molds."

No Staudacher aircraft is certified, and there are no plans to do so. Since every Staudacher is slightly different, rather than describe each one, we present a comprehensive review of the Staudacher 300GS designed for Mike Goulian to give you a flavor of Jon Staudacher's way of building his airplanes.

Staudacher 300GS

The Staudacher 300GS's fuselage is welded steel tube covered with fabric aft of the cockpit. Metal panels cover the fuselage forward of the cockpit up to the fiberglass engine cowling. The turtledeck is wood, delicately crafted and varnished inside, a fitting testimonial to Staudacher's nautical experience.

The wings are primarily wood with some composite reinforcement. Staudacher personally selects the Sitka spruce and sapele used in the wing construction. He inspects each piece of lumber visually and has ribbons cut off of them for further examination before he makes a commitment. Spiral grains are points of structural weakness to be avoided.

Staudacher 300GS

The spruce and sapele carry-through main wing spar is laminated with carbon fiber for extra strength. The wing ribs are also wood, and the wing is skinned with plywood covered with a layer of glass. The result is an extraordinarily strong, lightweight wing that has been static-loaded to plus/minus 23 Gs without failing.

In designing the aerodynamic characteristics of the high-lift wing, Staudacher made good use of state-of-the-art technology. The airfoil was developed on CAD (computer-aided design) equipment by Bob Lybeck, a leading aerodynamicist at McDonnell Douglas, who did a lot of work on the MD-11 program. It is also interesting to note that for the structural design of the wing, Staudacher relied entirely on racing-boat and wind-turbine blade structural technology.

All the control surfaces of the 300GS are covered with fabric. The control linkages are a combination of push rods and cable. The ailerons are mass balanced in a novel way. Instead of hefty counterweights, which could be stressed beyond limits and fail in high-G maneuvers, the leading edge of each aileron is filled with lead (buckshot), and the pivot hinges are buried within the aileron well aft of the leading edge. Light, high-strength, titanium shovels bolted to the aileron ease control loads. Aluminum springleaf gear struts custom-made by Staudacher complete the airframe.

Under the cowling is a 300-hp Lycoming IO 540 engine equipped with a Monty Barrett BPA cold-air induction sump, which boosts the power to about 320 hp. The power increase is achieved by routing the incoming air through an air plenum, which is separate from the oil sump (unlike conventional sumps). Thus the airflow remains unheated until it enters the cylinders.

Bolted to the nose is a three-bladed composite-coated wooden MT propeller built by Gerd Mühlbauer in Straubing, Germany, that is also used on the Extras and Sukhois.

The inverted system was custom-designed by Staudacher. It is yet another example of his preference for designing and building his own components instead of making do with off-the-shelf solutions. Just about the only items on the 300GS not made by Staudacher are the engine itself, the wheels and brakes, the Starlite canopy, the seatbelt, and a handful of instruments and avionics.

The end result of the 300GS design effort is a 320-hp monoplane with an empty weight of only 1280 pounds and a maximum gross weight of 1700 lbs, which yields a power loading of only 5.3 lbs/hp.

Mike Goulian flew a Staudacher 300GS to a second-place finish in the unlimited U.S. nationals in 1993 and also introduced Staudachers to world competition when he flew the airplane in the 1994 World Aerobatic Championships in Debrecen, Hungary.

300GS

SPECIFICATIONS

Engine	Lycoming AEIO-540
Horsepower	320
Propeller	MTV-9
Wingspan	24.4 ft
Wing area	107.8
Length	21.8 ft
Height	5.9 ft
Seats	1
Fuel capacity	42 gal

300GS

WEIGHTS AND LOADING

Gross weight	1700 lb
Empty weight	1.280 lb
Wing loading	15.7 lb/sq ft
Power loading	5.3 lb/hp
Limit load (pos/neg)	+12/–12

PERFORMANCE

Vne	267 kt
Maximum speed, sea level	178 kt
Rate of climb, sea level	3300 ft/min
Roll rate	320°/sec
Range	300 nm
Stall speed	48 kt
Takeoff ground roll	500 ft
Landing ground roll	900 ft

SUKHOI DESIGN BUREAU

The Sukhoi Design Bureau of the Soviet Union, traditionally a maker of military aircraft, including supersonic fighters, embarked on building unlimited aerobatic aircraft in the early 1980s. Its first effort, the Sukhoi 26, made its debut at the 1984 World Aerobatic Championship in Hungary. It appeared again in a more refined version in 1986 and went on to become the best Soviet unlimited aerobatic aircraft and a world-class competitor, reaping its share of medals in world competition.

Even before the Soviet Union collapsed and the Sukhoi Design Bureau became Russian, Brian Becker of Pompano Air Center, Pompano Beach, Florida, mounted a herculean effort to bring the SU-26 to the West. The airplanes caused a sensation at Oshkosh in 1989 when they arrived in a giant Antonov An-124 transport and wowed the crowds with a dazzling display of snaps and vertical rolls.

Working closely with the factory, Becker succeeded in having the airplane's finish quality, a key element of success in the Western market, greatly refined, and the airplane became a popular choice worldwide. It was soon joined by an equally compelling stablemate, the two-seat SU-29. In 1993, the lighter, more powerful Sukhoi SU-31 made its appearance as the successor to the SU-26.

SU-26MX

The SU-26MX is often described as exuding brute power, a perception created by its massive nose, which houses its 360-hp Vedeneyev M-14P radial engine. The engine swings a three-blade, constant-speed, wood-and-composite MT propeller that is also found on the Extras and CAPs. For pilots who fly behind Lycomings and Continentals, it is important to remember that the Vedeneyev swings the prop to the left, which requires the performance of certain maneuvers in the direction opposite to what they are accustomed. An unusual feature of the Vedeneyev's cooling control system is that instead of cowl flaps, an iris-like circular shutter behind the spinner expands and contracts to the pilot's command.

The airframe is of mixed construction. The fuselage is a welded stainless-steel tubular truss covered with a honeycomb composite skin. The engine cowling is aluminum. The exhaust system, the firewall, the landing-gear struts, and some minor components are titanium. The wing and empennage are composite with carbon spars. The wing also contains aluminum wing ribs. The ailerons and elevator are operated by pushrods, the rudder by cables.

The pilot's seat is exceptionally reclining, designed to put the pilot in a posture most resistant to G forces.

The starter system is unusual for pilots used to Western aircraft. It is an air-start system that utilizes onboard compressed air to swing the prop instead of an electrical starter motor. One advantage of this system is that it isn't affected by cold weather.

SU-29

The SU-29 was one of the first two-seat unlimited aerobatic aircraft capable of keeping up with the new generation of monster monoplanes. It is very similar to the SU-26, sharing many elements with it, including the engine. However, the airframe, including the wingspan is bigger. Two sets of shovels lighten the control loads on the massive full-span ailerons.

The SU-29 has one of the most generous canopies among aerobatic aircraft. It comes below shoulder level, and visibility from the cockpit is reminiscent of the view from the more modern fighters. With a full fuel load in the internal wing tanks, the SU-29 has more than five hours of endurance at economy cruise, a welcome capability on long ferry flights.

Sukhoi 26

SU-31

The SU-31 is the Sukhoi Design Bureau's latest entry into the unlimited aerobatic arena. It is an elegant airplane, a smaller, lighter and more powerful version of the Sukhoi 26, powered by a 360-hp version of the Vedeneyev engine. Its roll rate exceeds that of the SU-29.

First flown in world competition at the 1994 World Aerobatic Championships in Debrecen, Hungary, the SU-31 was still undergoing a shakedown period and was flown by pilots with little experience in type. Nevertheless, its pilots earned overall silver medals in both the men's and women's events, as well as a respectable share of medals in the individual events.

	SU-26	SU-29	SU-31
SPECIFICATIONS			
Engine	Vedeneyev M-14P	Vedeneyev M-14P	Vedeneyev M-14P
Horsepower	360	360	360
Propeller	MTV-3	MTV-3	MTV-3
Wingspan	25.6 ft	26.9 ft	25.6 ft
Wing area	138.8 sq ft	131.4 sq ft	138.8 sq ft
Length	22.4 ft	23.8 ft	22.4 ft
Height	7.5 ft	8.7 ft	7.9 ft
Seats	1	2	1
Fuel capacity	34 gal	72 gal	72 gal
WEIGHTS AND LOADING			
Gross weight	2315/1837 lb	2617/1900 lb	2425 lb
Empty weight	1450 lb	1647 lb	1543 lb
Wing loading	17.4 lb/sq ft	19.9 lb/sq ft	19.09 lb/sq ft
Power loading	6.1 lb/hp	7.2 lb/hp	6.7 lb/hp
Limit load (pos/neg)	+11/–9	+10.5/–9	+10/–9
PERFORMANCE			
Vne	243 kt	237 kt	243 kt
Maximum cruise speed	159 kt	143 kt	159 kt
Rate of climb, sea level	3540 ft/min	3189 ft/min	3540 ft/min
Roll rate	340°/sec	360°/sec	420°/sec
Range	432 nm	475 nm	475 nm
Stall speed	57 kt	56 kt	57 kt
Takeoff ground roll	197 ft	219 ft	360 ft
Landing ground roll	820 ft	838 ft	1131 ft

YAKOVLEV DESIGN BUREAU

The Yakovlev Design Bureau's YAK 55 has been around since 1982. Dissatisfaction with the initial wing led to a new wing design in 1984. The YAK 55 is the top unlimited aerobatic aircraft of a company whose creations dominated the aerobatic competition scene in past decades.

The YAK 55 has elegant lines, a midwing configuration, and is powered by the 300-hp Vedeneyev radial engine, Russia's staple powerplant for aerobatic aircraft. While it is an outstanding unlimited performer, it is not as competitive as its contemporaries. It is, however, very competitively priced, and might be a good economical choice for advanced pilots moving through the ranks.

YAK 55

SPECIFICATIONS

Engine	Vedeneyev M-14P
Horsepower	360
Propeller	V530TA-D35
Wingspan	9 meters
Wing area	14.8 sq m
Length	7.5 m
Seats	1

WEIGHTS AND LOADING

Gross weight	840 kg
Limit load (pos/neg)	+9/–6

PERFORMANCE

Maximum speed, sea level	450 km/h
Rate of climb, sea level	15 m/sec
Roll rate	450°/sec
Stall speed	100/105 km/h

ZIVKO

The Zivko Edge 540 is an excellent but little-known American 300-hp unlimited super-monoplane that is fully competitive with its contemporaries. It deserves much greater attention than it has received.

Zivko Aeronautics of Guthrie, Oklahoma, owned and run by Bill Zivko and his family, is an intriguing company serving a variety of niches in the aviation industry. Bill Zivko spent several years as the shop manager at Scaled Composites, Burt Rutan's famous skunk works, where he became an expert on composite construction before branching out on his own. Among other products, Zivko builds the composite airframes of the Perseus, an ultra-high altitude unmanned research aircraft that has a wingspan of 59 feet and can loiter at 80,000 to 100,000 feet for as long as four days.

With his experience in composites, it was natural for Bill Zivko to also branch out into aerobatics, and he got his start in the field by designing and building a superior composite replacement wing for Lasers, whose wooden wings were cracking with a disturbing regularity. From there, it was a logical step to develop his own 300-hp Laser-based design. The result is the Zivko Edge 540, available both as a kit or custom-built finished aircraft.

The Edge 540 has a meticulously crafted composite wing that has been static-loaded to 23 G without failing and a conventional steel tube fuselage. It has a midwing configuration and is one of the lightest of the super-monoplanes. It has the highest power loading among its contemporaries, which gives it a climb performance second to none. Its 420° per second roll rate is also an outstanding feature.

Since Zivko developed the replacement wing for the Lasers, it did not take much effort to also develop an in-house Laser fuselage, enabling Zivko to offer a four-cylinder aerobatic aircraft, the Edge 360, in kit or custom-built form. Though not competitive at the higher end, the Edge 360 is still a solid performer through the advanced category.

Edge 540

SPECIFICATIONS

Engine	Lycoming AEIO-540
Horsepower	327
Propeller	Hartzell three-blade
Wingspan	24 ft 4 in.
Length	20 ft 7 in.
Seats	1
Fuel capacity	52

Edge 540

WEIGHTS AND LOADING

Gross weight	1775/1527 lb
Empty weight	1170 lb
Power loading	4.7 lb/hp
Limit load (pos/neg)	+20/–20

PERFORMANCE

Vne	230 kt
Rate of climb, sea level	3700 ft/min
Roll rate	420°/sec
Stall speed	50 kt

ZLIN

The Czechoslovak Zlin 200, 300, and 500 series of aerobatic aircraft were the sensation of the 1960s. They were followed by the Zlin 50 series in the 1970s, and, as late as 1984, with Extras and Sukhois beating at the door, the top two men at the world aerobatic championships flew Zlin 50Ls. From then on, however, the aircraft has declined in competitiveness at the top level, and Zlin has not brought out a replacement aerobatic aircraft since.

The metal-and-fabric 300-hp Lycoming-powered Zlin 50L (the earlier versions were 260 hp) is worth mentioning because it remains an effective contender at the local level of the various categories of advanced aerobatic competitions. Its main attraction is that it can be had by aspiring advanced pilots for a price far below the cost of the more contemporary machines. It might be a cost-effective stepping stone on the way up in the ranks. While little known in the United States, it was always quite popular in Europe. One drawback is that it has a reputation for a slight disharmony in the control forces required to operate the ailerons and the rudders, which takes some getting used to.

Zlin 50L

Appendix A

Useful addresses

International Aerobatic Club
P.O. Box 3086
Oshkosh, WI 54903-3086
(414) 426-4800

Experimental Aircraft Association
3000 Poberezny Rd.
Oshkosh, WI 54903-3086
(414) 426-4800

Advanced Sukhoi Technologies
23A Policarpov St.
Moscow 125284 Russia

U.S. dealer:

Pompano Air Center
1401 Northeast 10th St.
Pompano Beach, FL 33060
(305) 943-6050

AkroTech Aviation (Giles 200, 202)
53774 Airport Rd.
Scrappoose, OR 97056
(503) 543-7960

AVIAT, Inc. (maker of the Pitts, Christen aircraft)
P.O. Box 1149
Afton, WY 83110
(307) 886-3151

Avions Mudry et Cie
BP 214 Aerodrome
27300 Bernay
France
33 32 43 47 34

U.S. dealer:

Mudry Aviation, Ltd.
Sr 1, Box 18T, #7
Bunnell, FL 32110
(904) 437-9700

Extra Flugzeugbau
Flugplatz Dinslaken
4224 Hunxe
Germany
Fax: 011 49 2858 7124

U.S. dealer:

®Aero Sport
P.O. Drawer 1989
St. Augustine, FL 32085
(904) 842-6230

One Design c/o Aircraft Spruce and Specialty
P.O. Box 424
Fullerton, CA 92632
(800) 824-1930

MT Propeller Entwicklung GmbH
P.O. Box 0720
Straubling D-94307
Germany 011 49 9429 8433

Staudacher Aircraft
2648 East Beaver Rd.
Kawkawlin, MI 48631
(517) 684-7230

Zivko Aeronautics (Edge 540)
502 Airport Rd.
Building 11
Guthrie, OK 73044
(405) 282-1330

Appendix B

Selected maneuver definitions, IAC Rule Book

In this appendix are examples of the definitions of maneuvers from the IAC Rule Book. You can see that the skills you acquire by learning to fly the building-block maneuvers of advanced aerobatics equip you to understand and be able to fly all the maneuvers defined in the Rule Book (and the FAI Catalog). Be aware that the rules may change from year to year, mostly in minor ways. Consult a current Rule Book when preparing for a competition. The definitions below are partial extracts.

Beginning and end of a figure

A figure begins at the moment the aircraft changes from a straight, wings level, horizontal flight path and ends at the moment the aircraft returns to a straight, wings level, horizontal flight path. The only exception to this are the exit lines on FAI Aerobatic Catalogue lines 7.7 and 7.8.

The aircraft shall fly in straight, wings level horizontal flight (upright or inverted) between figures.

The absence of a distinct straight, wings level horizontal flight path between figures will result in the grades being reduced by one (1) point in each case for each figure.

Lines

All lines are judged in relation to the horizon and the contest axes. For horizontal lines it is logical, with varying speeds and consequently varying angles of attack, to take the flightpath of the aircraft as the criterion (i.e., the competitor flies with the VSI indicating zero and the Judge must try to assess the line whilst ignoring the aircraft's attitude). The aircraft's heading must be parallel to the x or y axis, and the attitude must give a horizontal flight path all the time.

All lines that occur inside a figure have a beginning and an end which defines their length.

Whenever any kind of roll is placed on an interior line of a defined length, the lengths of the two parts of the line before and after the roll must be equal.

Loops and part loops

The loop is a figure from Family 7, but parts of them are in fact involved in many other families.

A loop must have, by definition, a constant radius, and be flown in the same vertical plane throughout. It starts and ends in a well defined line which, for a complete loop, will be horizontal. For a part loop, however, such lines may be in any other plane of flight, and will be defined by the aircraft's attitude.

The part loops of any one figure should all have the same radius. For example, a figure starts on a horizontal line, with a quarter loop, followed by a vertical line, and then another quarter loop. The quarter loops should have radii that are equal to each other and this means that the upper quarter loop must have a much slower angular velocity.

Rolling turns

These are simply rolls combined with various horizontal turns. Rolling turns may be started from upright or inverted flight. There are many factors to look for, and it is for this reason that you see very few in free sequences unless a competitor is forced to use them.

Combinations of lines

The transition from level flight to the 45° lines should be at a constant and reasonable ⅛ looping radius. All lines within the figure should be equal in length. The 90° transitions should have a constant and reasonable radius and not a square corner.

Rolls combined with tailslides

When rolls are combined with tailslides there must be an equal length of line before and after the roll(s). In the vertical downline the aircraft must attain a vertical attitude and project a downline before starting the roll(s).

Square loops and eight sided loops

The square and eight sided loops are actually to be flown as hesitation loops with lines of equal length and not as drawn in the catalog symbols. All lines, except the horizontal ones are judged on attitude.

All four lines in a square loop must be of equal length and be exactly parallel or at 90° to the main axis. The vertical lines are demonstrated by the aircraft's attitude; the other lines by flight path. The square loop is not corrected for wind other than for the length of the horizontal line, which must be wind corrected to keep the loop square. When rolls are flown on square loops, the roll or half roll must be centered on the horizontal or 45° line.

Vertical S

These figures are accomplished by half an inside loop and half an outside loop. Look for both half-loops to be of the same size and perfectly semicircular. The semi-circles should be placed one directly above the other except when a rolling figure is performed between them which will have the effect of displacing them by the length of the roll. No horizontal line is to be projected between the half loops. If any kind of roll occurs between them it must be executed in the horizontal, without a line being shown before or after.

Combination of lines, loops and rolls

This family of figures combines horizontal, 45° and 90° lines with ½, ¾, and ⅞ loops into some very complex figures.

As with all figures, there must be a clearly recognizable horizontal line at the start and finish of the figure. The initial horizontal line and the line at the end of the figure may be flown at different altitudes.

Whenever, inside a figure, a straight line connects loops and/or part loops and there is a rotation on that line, the rotation must be performed in the center of that line. The length of lines is not a criterion except that the part-lines before and after the roll must be equal.

Half loops that are flown after 90° and 45° lines still must have a constant radius, which requires a change of angular velocity during the half loop.

Rolls

Rolls may be performed on horizontal, 45°, or 90° lines; on complete loops; and between part loops and part loops/lines. They may be ¼, ½, ¾ and a full 360° in their rotation, up to two full rolls. Additionally, slow rolls may be flown in combination with turns as prescribed in Family 2 (rolling turns).

In all cases the same criteria apply: the rate of roll must be constant throughout the roll(s). The aircraft should continue to project, during the rolling portion, the prescribed plane and direction of flight.

Snap rolls

The main axis of the snap roll must be in the correct plane and direction of flight, which is recognized by attitude. However, the kind of movement around the main axis can differ between different aircraft type, and this should not affect the grading of the figure. Snap rolls have the same criteria as other rolls so far as the length of part-lines before and after the roll is concerned.

For all aircraft types the criteria for stopping the snap roll must be the same: that the attitude before starting the snap and the instant of stopping it must be identical and must correspond to the geometry of the figure.

The nose must definitely break the line of flight, indicating that a stall has occurred, or the figure earns a Zero. All snap rolls must be evaluated carefully to be sure that the competitor is not just aileroning the aircraft around. Aerobatic aircraft which have very rapid roll rates can occasionally fool a Judge in this figure.

Appendix C

Grading the Unlimited four-minute freestyle program

Compared to the voluminous collection of rules in the IAC Rule Book covering classic aerobatics, the guidelines for grading the four-minute freestyle program are quite loose and open to interpretation. The advantage of this approach is that it gives pilots maximum scope to be innovative in this program, the event many pilots consider to provide the most exciting opportunities for breaking new ground in advanced aerobatics. Here are the IAC rules governing the four-minute freestyle program.

This program will be graded by each Judge under four performance headings:

Originality	K=60
Versatility	K=60
Harmony and Rhythm	K=60
Execution	K=60
Total	K=240

These guidelines serve as a basis for defining these terms:

Originality

Entirely new figures, novel presentation of a familiar figure, unique handling of the aircraft, unique utilization of the aerobatic box and surprise factor.

Versatility

Using the total aircraft performance envelope and including a wide range of figure types, both catalogued and non-catalogued, such as gyroscopics.

Harmony and rhythm

Coherence of the flight, easy flow, continuous action, steady pace, choreography, dance quality and emotional stimulation.

Execution

Accurate flying, precision airmanship, wind utilization, engine and power handling, efficient energy usage, theatrical entertainment and audience retention.

Judges' worksheet

Judges can use whatever relative scale they prefer on a worksheet listing the competitor's flight and order number as they must base their individual preference under each performance heading relative to the preliminary grade they assigned to the first competitor. When all the competitors have flown, the Judge can scan the worksheet and see which competitors they gave the highest preliminary grade, for example, on originality. The judge will now assign a final grade to that competitor of a high value between 10 and 0, expressed in increments of 0.1 which will be recorded on the final grade sheet turned in for each competitor. The second highest preliminary grade for originality will be given a final grade below the highest, and so on for all competitors. Then in the same fashion, final grades can be given for each competitor on versatility, harmony and rhythm, and execution.

There are no positioning grades or boundary infringement penalties for this program. In addition, if the pilot incurs deadline penalties, "Execution" should also be downgraded.

Appendix D

International Aerobatic Club Achievement Awards Program

The IAC Achievements Awards Program was formulated to promote and advance sport aerobatics. The IAC sanctions many regional aerobatic contests every year, but at the same time, it recognizes that all pilots who fly aerobatics might or might not wish to enter competition but deserve recognition of their own abilities. Therefore, the IAC Aerobatic Achievements Award Program furnishes the mechanism in which competition and noncompetition pilots can work to reach various levels of proficiency. Aerobatic competitions benefit, because more people will be encouraged to enter, aerobatic education will be more widely disseminated, and finally, aerobatics as a sport will grow as more people learn of the enjoyment, fun, and comradeship that is known in aerobatic circles.

The IAC Aerobatic Achievement Awards program supplies a basis for which a pilot can increase proficiency in aerobatics and have a definite goal to work towards. These awards are not easy to achieve, and a high level of skill is required. They are, however, within the reach of every pilot. When the award is given, it is something to be proud of.

Moreover, many people can now be introduced to aerobatics with this incentive to achieve a high degree of skill under a program of strictly controlled conditions, that are tightly monitored with an eye always on safety.

GENERAL RULES

A. The flight or flights made to obtain said award or awards shall satisfy the definitions of the following regulations.

B. The awards are international in scope, as is the IAC. All pilots from all countries may achieve the awards as long as they meet the qualifications and requirements of this program.

C. An applicant for an IAC Aerobatic Achievement Award must be a member in good standing of the International Aerobatic Club, Inc.

D. All IAC Official Contest Rules will prevail with the following exceptions:

1. There shall be no maximum altitude except that the figures must be easily seen from the ground with the naked eye.

2. Technical inspections will not be required.

CATEGORY CLASSIFICATION

A. There are five (5) categories of achievement awards that can be earned. They are a gradual step up in the proficiency and skill required to successfully complete the figures in each respective category. These five categories are:

1. Basic

2. Sportsman

3. Intermediate

4. Advanced

5. Unlimited

B. Each category has a designated set of figures which must be completed under the following criteria. See the following list of figures. These awards may be earned as follows:

1. The required figures can be completed in one flight or individually on different days. When the application is mailed in, however, all the figures shall be judged to have been completed.

2. An approved IAC judge must be in attendance and certify completion of each figure and category. The judge will sign the application for the award and list his or her IAC number.

3. A grade of five (5) on a scale of 1–10 will be considerd the minimum for each figure. Chapter VIII (of the IAC Rule Book) will furnish the necessary criteria for these figures. There will be no overall score.

4. Minimum altitudes for each category will prevail at all times. Breaking of minimum altitude limits will result in a grade of zero (0) for each figure performed under the lower altitude limit.

The minimum altitudes are:

a. Basic   1,500 ft

b. Sportsman   1,500 ft

c. Intermediate   1,200 ft

d. Advanced   800 ft

e. Unlimited   328 ft

5. In cases of duplicate figures or subsequent awards, the figure(s) must be re-flown for each award. All figures listed for each respective category must be flown in order to qualify for the award.

6. To perform aerobatics below 1,500 ft requires waivered airspace.

7. An applicant for an award must be the sole occupant of the aircraft while qualifying for said award.

8. For the regular award (without stars), the aerobatic zone, as outlined in Chapter VI, will not be observed.

C. In addition, awards with "stars" may be earned in all categories. The awards with "stars" will be governed by the following rules:

1. All categories with "stars" must be done in competition at a sanctioned IAC contest with all of the "IAC Official Contest Rules" applying.

2. The applicant will be graded on each figure on all of the flights which are completed at the contest. The Unlimited Four Minute program is not to be included as a flight.

3. The figures required for the Achievement Awards with "stars" will be the current sequences for the particular category as set forth by the IAC for that year.

4. A minimum raw grade of five (5) for each figure and positioning (excluding the low grade by figure) is required in a contest with four or more judges. In a three judge contest all grades must be five (5) or more.

AWARDS

A. A certificate suitable for framing and a wallet card with a permanent number will be issued when a particular category has been satisfactorily completed, attested to and sent in.

B. Distinctive patches and badges will be available to the pilots who have qualified for these respective awards.

C. There will actually be ten (10) separate lists with their own numbering systems. Note that it will be possible for a pilot to appear on the various lists ten (10) different times. The Achievement Awards Chairman will keep these records and maintain the permanent lists.

D. The application for Achievement Awards can be obtained from the Achievement Awards Chairman listed in SPORT AEROBATICS or from the Registrar at a sanctioned IAC contest.

CATEGORY FIGURE LISTS

A. Basic
1. Loop
2. Spin (one turn)
3. Roll (any type)

B. Sportsman
1. Spin (one turn)
2. Loop
3. Immelmann
4. Half Cuban

5. Snap roll

6. Slow roll

7. Hammerhead

8. Reverse half Cuban

C. Intermediate

   1. Spin (one turn)

   2. Immelmann

   3. Half Cuban

   4. Square loop

   5. 45° diving snap roll

   6. 45° climbing slow roll

   7. Hammerhead with quarter rolls up and down

   8. Reverse half Cuban

   9. 4 point hesitation roll

D. Advanced

   1. Spin (one turn cross-over)

   2. Immelmann

   3. Half Cuban

   4. Square loop

   5. 45° diving or climbing snap roll

   6. 8 point hesitation roll

   7. Reverse half Cuban

   8. Outside loop

   9. English bunt

   10. Cuban (inside-outside)

   11. 360 rolling circle (number of rolls optional)

   12. Vertical half roll with push over

E. Unlimited

   1. Hammerhead with inverted entry

   2. Inverted spin (one turn)

   3. 360° rolling circle with rolls to outside (number of rolls optional)

   4. Tailslide

   5. Cuban (double outside)

   6. Outside snap roll

   7. Vertical diving snap roll (inside or outside)

   8. 8 point hesitation roll

   9. Outside loop

   10. Vertical roll (of at least 180 rotation)

   11. 8 sided loop

Appendix E

Freestyle program composition forms

Following are the IAC freestyle program composition forms. The judge's score-sheet (Fig. E-1) must contain all the maneuvers and associated information. This form is used by the judges to score your freestyle performance. The other forms (Figs. E-2 and E-3) are blank sequence cards.

A	INTERNATIONAL AEROBATIC CLUB SCORESHEET				CATEGORY:								pilot's number	
No.	symbol	catalogue no.	K	total K	grade	remarks	No.	symbol	catalogue no.	K	total K	grade	remarks	
1							13							
2							14							
3							15							
4							16							
5							17							
6							18							
7							POSITIONING							
8							FIGURE TOTAL K = _____							
9							INCLUDING POSITIONING = _____							
10							JUDGE'S NAME _____							
11							JUDGE'S NUMBER _____							
11							FREE PROGRAM CHECK BY:							
12							_____							
							(signature/date)							
							AIRCRAFT TYPE _____							

IAC

Fig. E-1. Judge's scoresheet.

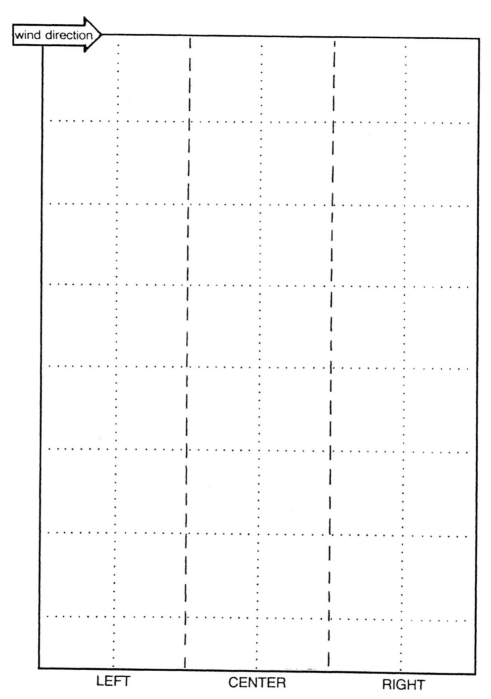

Fig. E-2. Blank sequence card.

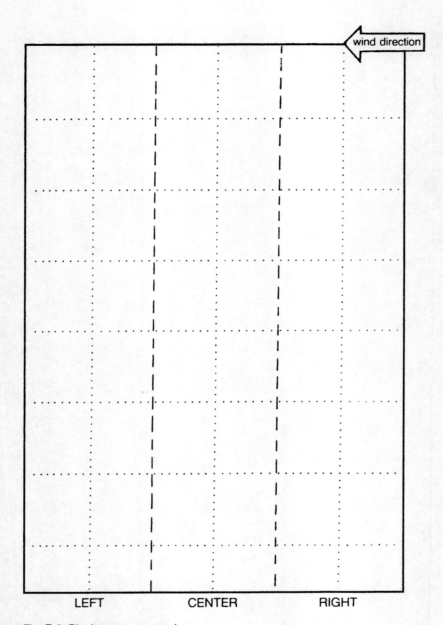

Fig. E-3. Blank sequence card.

Appendix F

Sample page from FAI Aerobatic Catalog

7. LOOPS AND EIGHTS

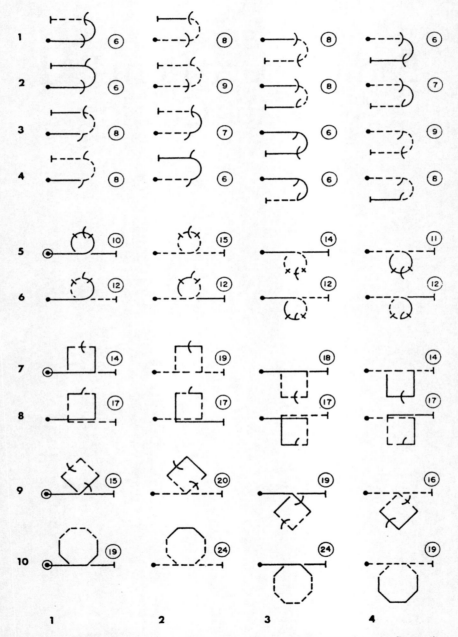

Fig. F-1. A sample page from the FAI Aerobatic Catalog. The commas on the figures indicate where rolls may be inserted.

Appendix G

Aerobatics schools

This appendix contains information on aerobatic schools in the United States that offer advanced aerobatic instruction. While aerobatic schools may come and go, like all flight schools, experience has shown that many of these schools have been around for a long time and are likely to remain in operation for a long time to come.

Several of these schools offer services in addition to advanced aerobatics and have aircraft other than advanced aerobatic aircraft. Only information pertinent to advanced aerobatics is included, however. Most schools provide instruction in your own two-seat aircraft and several will critique your competition flying. Try the schools listed here, and contact the IAC for a current list of aerobatic schools.

Note that we do not endorse any of the schools listed or the qualifications, experience, or expertise of their operators.

Addison Flight Training Center
Contact: Henry Goff Undo Schmidt-Sinns
4545 Eddie Rickenbacker Drive
Dallas, TX 75248
(214) 991-1529 (214) 891-8020
Type aircraft used: Pitts S-2B
Type advanced instruction: competition customer designed

Aerobatics by Maltaire
Contact: Rich Goldstein
Hering Drive Long Island
McArthur Airport
Ronkonkoma, NY 11779
(516) 471-2270 (office) (516) 471-6711 (fax)
Type aircraft used: Cap 10B
Type advanced instruction: through unlimited

AeroTech Enterprises, Inc.
Contact: Marilyn Hubbard
Allaire Airport
P.O. Box 2393
Farmingdale, NJ 07727
(908) 477-3345 (home)
Type aircraft used: Pitts S-2B
Type advanced instruction: intermediate, advanced competition

Aragon Aviation, Inc.
Contact: Cecilia Aragon
1227 The Alameda Berkeley, CA 94709
(510) 527-4466
Type aircraft used: Pitts S-2B
Type advanced instruction: advanced customer designed

Aurora Aviation
Contact: Ralph Riddell
22785 Airport Road
P.O. Box 127 Aurora, OR 97002
(503) 222-1754 (office)
(206) 896-1160 (home)
Type aircraft used: Pitts S-2B
Type advanced instruction: custom designed

Aviation Unlimited
Contact: Richard Crofton-Sleigh
872 Clayton Street
San Francisco, CA 94117
(Aircraft at Hayward Airport)
(510) 785-0800 (office)
(415) 731-7697 (home)
Type aircraft used: Pitts S-2A, Pitts S-2B
Type advanced instruction: through unlimited custom designed

Central Ohio Aerobatics
Contact: Ken and Barb Hadden
943 Mulberry Drive Columbus, OH 43235
(614) 885-5685
Type aircraft used: Pitts S-2B
Type advanced instruction: competition and critiquing custom designed

Chandler Air Service, Inc.
Contact: John Walkup
1675 E. Ryan Road
Chandler, Az 85249
(602) 963-6420
Type aircraft used: Pitts S-2B
Type advanced instruction: competition customer designed

CP Aviation
Contact: Chris Thoms
830 E. Santa Maria St #301
Santa Paula, CA 93060
(805) 525-2138 (office)
(805) 933-3865 (fax)
Type aircraft used: Pitts S-2A
Type advanced instruction: customer designed

Delat Aviation
Contact: Fred DeLacerda
1802 W. Wright Stillwater Airport
Stillwater, OK 74075
(405) 624-0719 (office)
(405) 624-0955 (home)
Type aircraft used: Super Decathlon
Type advanced instruction: through intermediate custom designed

Dent Air
Contact: Bill Finagin
6 Romar Drive
Annapolis, MD 21403
(410) 956-0047 (hangar)
(410) 263-2740 (home)
(410) 263-4693 (fax)
Type aircraft used: Pitts S-2B
Type advanced instruction: customer designed

Everglades Aerobatics
Contact: Bruce Thalheimer
3644 Boca Ciega Drive
Naples, FL 33962
(800) 729-8994 (office)
(818) 744-4666 (office)
(818) 744-4773 (home)
Type aircraft used: Pitts S-2B
Type advanced instruction: customer designed

Executive Flyers Aviation
Contact: Mike Goulian
Terminal Building
Hanscom Field
Bedford, MA 01730
http:www.executiveflyers.com
(617) 274-7227 (office)
(617) 274-6719 (fax)
Type aircraft used: Pitts S-2B, Extra 300L
Type advanced instruction: through unlimited custom designed

Flugvergnuegen
Contact: Dagmar Kress
1020 N.W. 62nd Street
Ft. Lauderdale, FL 33308
(305) 566-4508
Type aircraft used: Pitts S-2B Extra 300L
Type advanced instruction: competition customer designed

Fly for Fun
Contact: William H. Thomas
400 E. Airport Avenue
Venice, FL 34285
(813) 484-8183 (office)
(813) 493-5896 (home)

Summer:

71 S. Shore Drive
Cuba, NY 14727
(716) 968-3758
Type aircraft used: Pitts S-2B
Type advanced instruction: through advanced

Great Planes Aerobatics
Contact: John C. Morrissey
245 N.E. Edgewater Drive
Lee's Summit, MO 64064
(816) 373-8675
Type aircraft used: Pitts S-2A and customer's
Type advanced instruction: through unlimited customer designed critiquing training
camps low level aerobatics

Harvey & Rhin Aviation, Inc.
Contact: Debby Rhin or Dr. Eoin Harvey
101 Airport Blvd.
LaPorte, TX 77571
(713) 471-1675 (office)
(713) 473-2983 (fax)
Type aircraft used: Pitts S-2A, Pitts S-2B
Type advanced instruction: through unlimited

Lenair Aerobatics
Contact: Michael C. Church
19531 Campus Drive, Suite 7
Santa Ana, CA 92707
(714) 852-8850 (office)
(714) 548-1713 (home)
Type aircraft used: Pitts S-2A, CAP 10B
Type advanced instruction: through intermediate customer designed

Littlefield, Gene
Contact: Gene Littlefield
23657 DuPage CC Drive
Plainfield, IL 60544
(815) 436-5590
(815) 436-8280 (fax)
Type aircraft used: Decathlon and customer's
Type advanced instruction: through advanced customer designed

Mass Acro Aerobatic Center
Contact: Jim Thompson
Box 6
Plymouth Airport
Plymouth, MA 02360
(508) 747-1719 (office)
(617) 585-7963 (home)
Type aircraft used: Pitts S-2B
Type advanced instruction: through unlimited

Mudry Aviation, Ltd. (home of the French Connection)
Contact: Montaine Mallet Flagler
County Airport
SR1, Box 18T, #7
Bunnell, FL 32110
(904) 437-9700 (office)
(904) 437-1177 (fax)
Type aircraft used: CAP 10B
Type advanced instruction: customer designed

New Attitude Aerobatics
Contact: John Blum
10860 W. Evans, #50
Boulder, CO 80227
(303) 980-5667
Type aircraft used: Pitts S-2B
Type advanced instruction: through intermediate customer designed

Pietsch Flying Service
Contact: Warren Pietsch
#14 Minot International Airport
Minot, ND 58701
(701) 852-4092
Type aircraft used: Pitts S-2B
Type advanced instruction: customer designed

Pompano Air Center
Contact: Brian Becker
1401 N.E. 10th Street
Pompano Beach, FL 33060
(305) 943-6050 (office)
(305) 943-0829 (fax)
Type aircraft used: Pitts S-2B
Type advanced instruction: through unlimited customer designed

Rocky Mountain Aerobatics
Contact: John Szakach
8 S. Evanston Way
Aurora, CO 80012
(303) 363-9817 (home)
Type aircraft used: Pitts S-2B
Type advanced instruction: customer designed

Sunair Aviation Center
Contact: Gil Monti, Greg Loughran
15115 Aitport Drive
Scottsdale, AZ 85260
(602) 991-0611 (office)
(602) 943-0240 (home)
Type aircraft used: Pitts, S-2B
Type advanced instruction: customer designed

Thomason Aerobatic Center
Contact: Randy Gagne
16700 Roscoe Blvd.
Van Nuys, CA 91406
(818) 908-8267 (office)
(818) 908-5800 (office)
(818) 908-5807 (fax)
Type aircraft used: Pitts S-2B
Type advanced instruction: through unlimited customer designed

US Wings
Contact: Michele Thonney
2501 S.E. Aviation Way
Stuart, FL 34996
(407) 770-2222 (office)
(407) 335-3365 (fax)
(407) 335-3364 (home)
Type aircraft used: Pitts S-2B
Type advanced instruction: through advanced competition

Unusual Attitudes
Contact: Bret Ebaugh
7755 S. Peoria, Hangar 5
Centennial Airport
Englewood, CO 80112
(800) 43-9901
(303) 688-2233 (home)
Type aircraft used: Pitts S-2B
Type advanced instruction: custom designed

Williams Aviation, Inc.
Contact: Ray Williams
P.O. Box 397
Springfield, TN 37172
(615) 255-9299
(615) 384-9388
Type aircraft used: Pitts S-2B
Type advanced instruction: customer designed

Worley Aviation
Contact: A. C. "Jim" Worley
3012 W. Broad Street
Richmond, VA 23230

Airport:

6920 Terminal Road
Quinton, VA 23141
(804) 353-2641 (office)
(804) 932-4408 (home)
(804) 932-8308 (airport)
(804) 932-3984 (airport)
Type aircraft used: Pitts S-2A
Type advanced instruction: customer designed

Index

ABOUT THE AUTHORS

Geza Szurovy is an experienced pilot with instrument, multiengine, DC-3, and glider ratings. He is an aviation photojournalist and the award-winning author of such aviation books as *Learjets*, and *Basic Aerobatics* (co-authored with Mike Goulian).

Michael Goulian was the 1995 Unlimited U.S. national aerobatic champion, the 1992 winner of the prestigious Fond du Lac Cup, and is a member of the US Aerobatic Team. Mike holds an Airline Transport Pilot rating and is an aerobatic flight school operator and instructor. He is also a well-known airshow pilot.